Proceedings

46 Congreso Internacional de Americanistas
International Congress of Americanists

Amsterdam, Netherlands 1988

New Directions in American Archaeoastronomy

edited by
Anthony F. Aveni

BAR International Series 454
1988

B.A.R.

5, Centremead, Osney Mead, Oxford OX2 0DQ, England.

BAR -S454, 1988: 'New Directions in American Archaeoastronomy'

ISBN 9780860545835 paperback
ISBN 9781407347394 e-book
DOI https://doi.org/10.30861/9780860545835
A catalogue record for this book is available from the British Library
This book is available at www.barpublishing.com

CONTENTS

LIST OF CONTRIBUTORS

Anthony F. Aveni, Russell B. Colgate Professor of Astronomy & Anthropology, Colgate University, Hamilton, New York 13346, USA

Victoria Bricker, Professor of Anthropology, Tulane University, New Orleans, Louisiana 70118, USA

Clemency Chase Coggins, Peabody Museum of Archaeology & Ethnology, Harvard University, Cambridge, Massachusetts 02138, USA

R. David Drucker, Geochron Enterprises, Cambridge, Massachusetts 02138, USA

Horst Hartung, Faculty of Architecture, University of Guadalajara, Mexico

Susan Milbrath, Florida State Museum, University of Florida, Gainesville, Florida 32611, USA

Sebastián Robiou-Lamarche, Aptdo. Postal 1626, San Juan, P.R. 00903, USA

Franz Tichy, Professor für Geographie, Universität Erlangen-Nürnberg, Kochstr. 4, D 8520 Erlangen, RFA

M. Jane Young, Associate Professor of American Studies, University of New Mexico, Albuquerque, New Mexico 87131, USA

Michael Zeilik, Professor of Astronomy, University of New Mexico, Albuquerque, New Mexico 87131, USA

R. Tom Zuidema, Professor of Anthropology, University of Illinois, Urbana, Illinois 61801, USA

INTRODUCTION

These papers are among some three dozen from twelve countries read at a joint symposium entitled "Etnoastronomia y Arqueoastronomia en las Americas", chaired by Edmundo Magaña and Anthony F. Aveni, and held at the 46th International Congress of Americanists in Amsterdam, July 1988. When collection and editing of the full complement of papers proved unwieldy, it was judged to be more realistic to undertake to publish the archaeoastronomy papers separately; thus the present volume. As was the case with a similar symposium held at the 44th ICA in Manchester in 1982 (B.A.R. 174, 1983), the editor's principal goal has been to disseminate results of current research in the growing interdisciplinary field of archaeoastronomy both expediently and efficiently to our interested colleagues.

Most of the contributors to this volume will be familiar to those who have read earlier collected works on archaeoastronomy in the Americas; however, the careful reader will note the growth and advancement of the interdiscipline indicated by the increased depth and breadth of inquiry undertaken by each of the authors into problems they had already dealt with to some extent in earlier publications. Thus, Zuidema has more finely honed his arguments concering the placement of the sun pillars on the horizon of Inca Cuzco, Aveni and Hartung have re-examined their earlier studies of the placement and orientation of the major temples of Tikal in the light of recently acquired knowledge concerning the dynastic sequence that ruled the great Maya city, and Bricker has begun to bring together the edges of the hazy gap between written Maya astronomical ephemerides and almanacs. The arguments of Coggins & Drucker and Milbrath concerning the orientation of buildings at Dzibilchaltun and Chichén Itzá, respectively, are set upon the more solid foundations of the bases of evidence provided by the Maya written and iconographic record.

Three of the nine papers in the collection address broader questions that transcend the study of single sites and perhaps even individual culture groups. Presentations by Zeilik and Young on the native American people of the US southwest and the paper by Tichy on Mesoamerica inquire into the rhythmic nature of the calendar, concepts of directionality and notions of angular measurement in general. The appearance of such overview papers in a proceedings volume reflects the degree to which contributing scholars are thinking well beyond the more narrow, individualized research approaches that characterized our interdiscipline a generation ago. Archaeoastronomy has indeed set upon new directions that carry it well beyond alignment hunting!

While Mesoamerica dominates the present volume (four papers specifically address Maya astronomy), we are pleased to see important contributions representing South America and the U.S. southwest appearing under the same roof. And we are especially grateful to Robiou-Lamarche for submitting his work on the Antilles, a new astronomical horizon for most students of archaeoastronomy.

The editor is grateful to Lorraine R. Aveni and Jim McCoy for their technical assistance in the presentation of this collection.

Anthony F. Aveni
1 November 1988

"ARCHAEOASTRONOMY AND DYNASTIC HISTORY AT TIKAL"

A. Aveni & H. Hartung

In the 1970's, Hartung (1975, 1977) proposed a number of logical and structural hypotheses by which deliberate orientation and planning schemes, particularly of an astronomical nature, may have been incorporated in Mesoamerican ceremonial centers. Over the next two decades he and Aveni surveyed large numbers of Mesoamerican archaeological sites with a view toward testing and applying some of these hypotheses. These studies helped to develop the broader discipline of archaeoastronomy, which focuses on understanding the nature of ancient astronomical practice utilizing both the written and the unwritten record. More recently, Aveni & Hartung (1986) have reviewed Maya city planning as it pertains to functional orientations that relate to the establishment of environmental calendars, wherein the movement of celestial bodies at the horizon, particularly the sun and Venus, served as timing devices, thereby indicating where certain structures ought to be placed and how they should be oriented.

A portion of one of these studies (Hartung 1977, pp 112-118) deals with the ruins of Tikal. The data were based initially upon the form and arrangement of buildings gleaned from a study of the Tikal map (Carr & Hazard 1961). Though we had only little knowledge of Tikal chronology at that time, our early studies nevertheless revealed the existence of several elements that suggest intentional planning. Alignment data were collected with surveying equipment by the authors at the site in January 1975. They were supplemented by further precise measurements made in January 1978 and January 1987.

In the past decade-and-a-half, much inscriptional and archaeological evidence has come to light regarding the dynastic history of Tikal and we now have a much clearer, though not yet fully comprehensive, idea of the chronology of the building complexes and even of some of the individual buildings. Moreover, we know and in some instances can make educated guesses about which members of the royal line are buried beneath the different structures. For example publications by Coggins (n.d.), Jones (1977), and Jones & Satterthwaite (1982), stress the valuable historical interpretations of the hieroglyphs at Tikal.[1] These data have stimulated further studies on Tikal, and in particular inquiries relating to the placement of buildings.

Based on the latest inscriptional and archaeological evidence, the building sequence, as best we can lay it out, is given in Table 1.

TABLE 1. BUILDING SEQUENCE AT TIKAL

```
No. Acropolis --> 5D33-1st, TII & TI
5D33-3rd & 2nd        probably conceived as a
                      commemorative pair

    V^A <-- IV <-- VI <-- III^B <-┘
 (or IV <-- V)
```

Notes to Table:

A. But see Coe (1967) who has I and V contemporary (with IV following about ½ century later) because I was finished after the death of A, while at the same time the initiation of V could have taken place.

B. Coe (1967) says that III follows I by 100 years, based on dates on St. 24. If A started building I before his death (in AD 722 or 734), then III could have been built 100 years later, based on the St. 24 dates.

In this paper we re-examine our earlier ideas on the role of astronomy and geometry in the planning of Tikal, giving particular attention to the new data concerning the genealogical arguments. The major portion of such arguments deals with the construction by the dynasty starting with Ruler A (nick-named Ah Cacau) at the end of the seventh century.

The dominant (approximately 10° E of N) orientation at Tikal is embedded in the North Acropolis: it is the same orientation as that of Temples I and II (Figs. 1 & 2), which confront one another immediately to the south and delimit the Great Plaza on the east and west. Temples IV and VI (although facing west) also have the same approximate orientation and the north front of Temple V deviates only a little more than 10 degrees E of N. Only Temple III is different from the others, lying 18°16' off cardinal north to the east. No convincing explanation for these deviations, astronomical or otherwise, has been offered since their peculiarity was first pointed out by Tozzer (1911: 106), but we believe the solar zenith passages at Tikal may have played a role (see summary in Table 2).

The six large pyramids (Temples I-VI) were built over a little more than one hundred years by the dynasty that began with Ruler A. An inauguration date on Lintel 3 of Temple I reads 9.12.9.17.16 5 Cib 14 Zotz (AD 682) on F8-E9, preceding the glyph of "seating." At F10-F11 the names of Ruler A and the Tikal emblem glyph are given. Burial 116, the tomb in Temple I, contains the remains of this ruler, while Temple II may have been dedicated to his wife (Coggins n.d.: 455), though no burial has been found there. One of the original five beams or wooden lintels of the middle doorway of this temple shows a person, who may represent this woman.

Fig. 1: Temple I viewed from the doorway of Temple II (Photo by Hartung).

Fig. 2: View from the doorway of Temple I to Temple II (foreground) and to Temples III and IV (background) (Photo by Hartung).

Recently, M. Miller (1985) has examined the spatial organization of the Late Classic Tikal pyramids in terms of the structure of the ruling family. She interprets TI and TII to be personifications of husband and wife, the female being "shorter, squatter, and broader than the male" (p. 8). She also refers to the dramatic breaking, by the east-west orientation of TI and TII, of the pattern of the structures built on the North Acropolis by the Early Classical rulers as the concrete image of a new order (p. 8).

Since the beginning of Tikal (around 600 BC), the North Acropolis, with its particular funerary architecture, had been the sacred burial ground of the Tikal rulers. But near the end of the Early Classic Period signs of change began to occur. Some scholars have argued that a cyclical rule of succession gave way to a dynastic political system.[2] This transition is said to have occurred with Curl Nose (AD 379-426), a possible usurper who may have come from the south, and his son Stormy Sky (AD 426-457).

The tomb of the first of the new dynasty rulers may be Burial 48 at the southern limit of the North Acropolis and on its axis (but see Note 1). A small temple (Structure 5D-33-3rd) was built above it, which later, in the middle of the fifth century, was covered by 5D-33-2nd, possibly constructed by Curl Nose's son. This structure represents the most ornate example of Early Classic Tikal funerary architecture and it persisted largely unchanged until the end of the seventh century, when it was partially dismantled and covered by the conspicuous, tall pyramid 5D-33-1st.

Apparently Ruler A erected 5D-33-1st as his first major building after taking power at Tikal in AD 682. The intention of this building was two-fold: first, to reaffirm his relation to his ancestors Curl Nose and Stormy Sky (Coggins n.d.: Table 4; Jones 1977: 62) and second, to commence a new kind of architectural expression suitable to his dynasty.

Meanwhile, structure 5D-33-2nd was relatively low (about 12 m high including the "cresteria"); its facade was accentuated by rectangles enclosing stucco polymorphic reliefs. The seventh century pyramid, 5D-33-1st, anticipates the form of Temples II and I that would follow it (particularly the latter with its vertical accent).

We consider Structure 5D-33-1st to be of vital importance in the planning scheme of Tikal. It embodies a link between the architectural past and the revolutionary ideas of the future dynasty. Tradition seems stronger at this location, where the tomb of the founder of the dynasty is fixed on axis with the temples of the North Acropolis. Also, Stela 31 representing Stormy Sky, which may have been positioned in front of 5D-33-2nd, was placed inside the temple of 5D-33-2nd when Ruler A covered this structure over with 5D-33-1st.

It is important to realize that today, the larger part of 5D-33-1st is no longer standing; thus the view is left open from the Great Plaza to the North Acropolis[3] (Fig. 3). The earlier building of Ruler A had previously closed off this view, thus lending importance to the future spatial relations between the Great Plaza and Temples I and II.

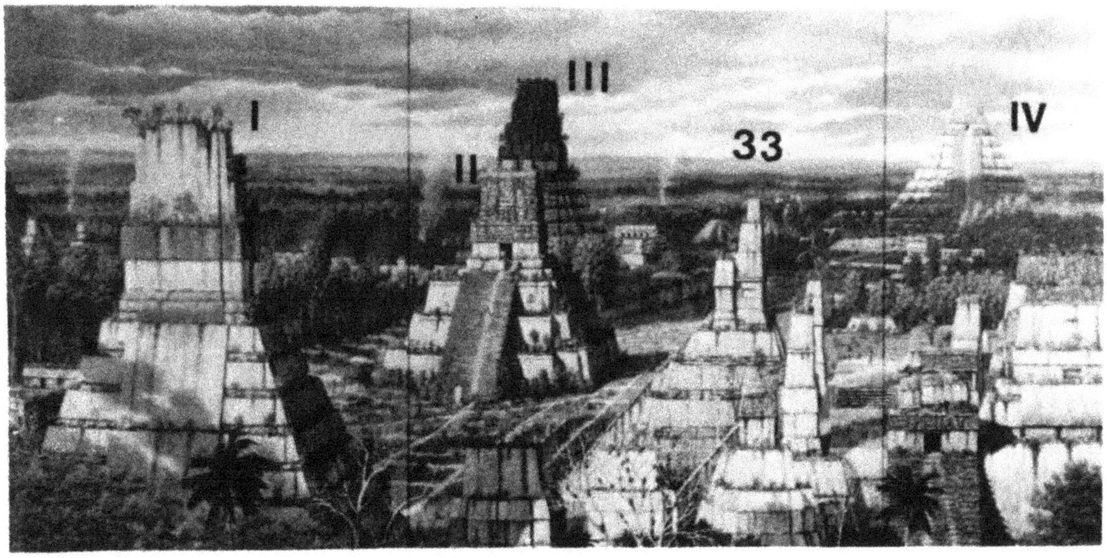

Fig. 3: Tikal diorama viewed from the east, adapted from "An Artist's Conception of Tikal at AD 1000" in A. Miller (1986) (with permission).

Judging from the Tikal map of 1961, a line between the doorways of Temples I and II lies at 90° to the approximate N-S axis of the North Acropolis. The distance from the crossing point of both lines to Temple II is somewhat larger than to Temple I, which may be due to an alignment reference of Temple II to the central pyramid (5D-104) of the South Acropolis, an earlier construction.

Archaeological data suggest that Temple II was constructed (around 700 AD) before Temple I, where Ruler A was buried. If, as it seems, the wooden lintels of Temple I were elaborated in the lifetime of Ruler A, one can suppose that he also prepared the construction of the pyramid and even may have initiated it.[4]

Ruler A also initiated a new type of building: the Twin-Pyramid Complex. The first of these, Complex M (or 3D-1), is situated exactly north of of 5D-33-1st, while the second, Complex N (or 5C-1), is located exactly to the west, each of these directions deviating noticeably from

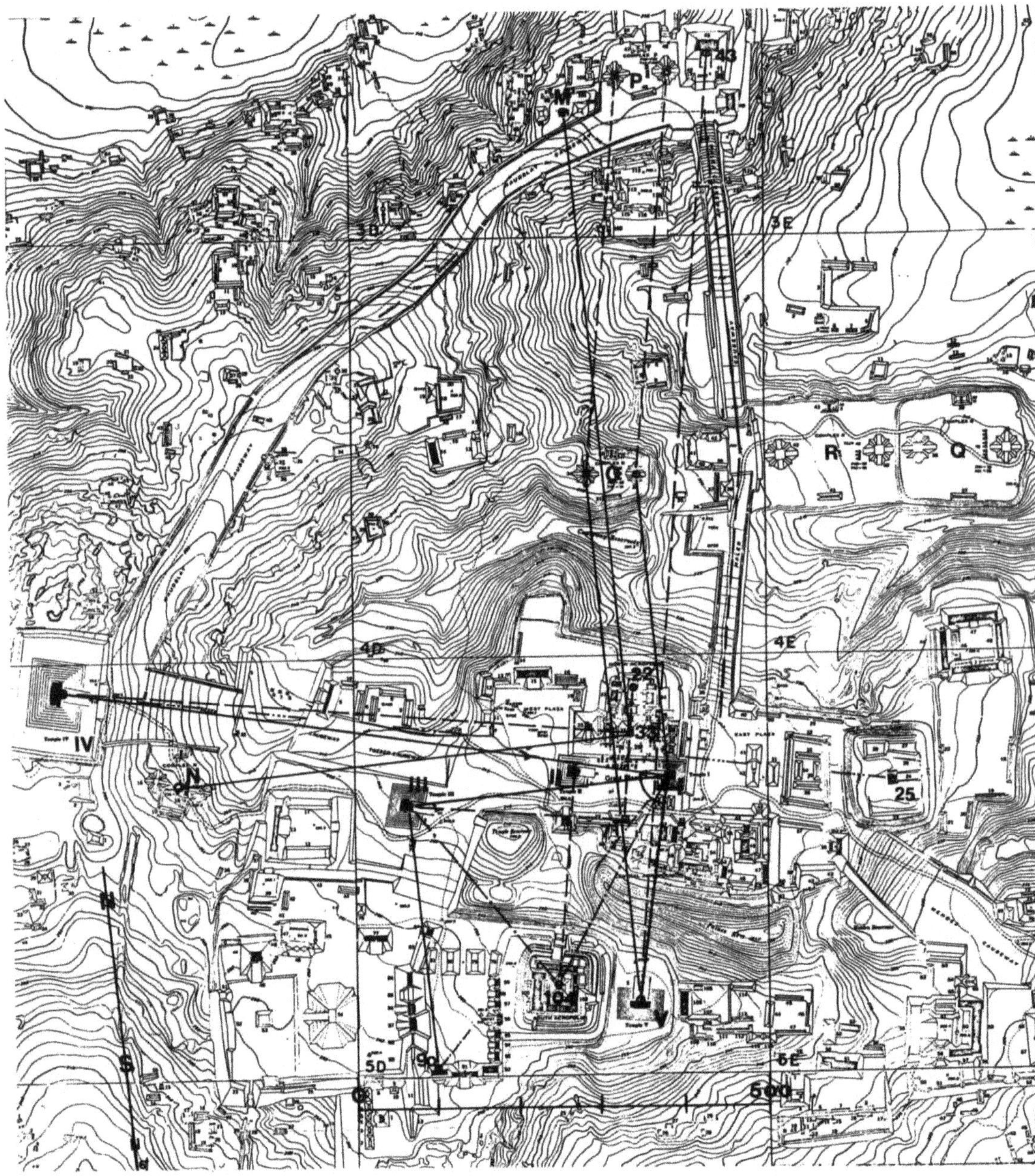

Fig. 4: The Tikal Map (Carr & Hazard 1961) showing reference lines and scale added by the authors. Vertical grid lines fixed to magnetic north, which deviates approximately 5½° to the east of astronomical north.

the dominant axis of the North Acropolis, as can be seen in Fig. 4. Cases of right angles between two notable directions such as these can be detected in other examples as well.

In placing both complexes, the reference was taken from the center of the Great Plaza. The visibility of Complex N from 5D-33-1st seems to have been important and the visual line taken from the earlier Complex M and directed toward the South also appears to be decisive. It should be noted that in Complex M we find Altar 14 with the katun-ending date 9.13.0.0.0 8 Ahau 8 Uo and Stela 30 with the earliest representation of Ruler A. Complex N has the elaborate Stela 16 and Altar 5. As if to accentuate the north-south axis, the sides (and the axis) of the northern part of the Maler Causeway (leading to the North Zone) also are directed exactly to the north (see right-most solid line in Fig. 4). The southern segment of the causeway (dashed line) is directed parallel to the axis of the North Acropolis. The two parts join at a point where the twin-pyramid complexes Q and R extend to the east, and Complex O to the west.[5]

The later Complex O has by seriation an assigned date of 9.15.0.0.0 = AD 731. The monuments in the North enclosure (Stelae P55 and altar P47) are plain. If one accepts the last date on an offering in Burial 116 as the day of the death of Ruler A, then he would have died in 9.14.11.17.3 = AD 722. There could have followed a time of transition until his son, Ruler B, took office in AD 734 (his inauguration date appears on Stela 21 in front of Temple VI, more than 1200 meters to the southeast of the Great Plaza). Between these two dates 12 years lapsed, during which a katun (9.15.0.0.0 = AD 731) was completed; thus the necessity to build a complex.[6]

If Ruler A had died in 722, his Temple I surely would have been completed by 731, and even if it was not, the position of the entrance to the temple should have been decided, in order to define, exactly toward true north, the center-point of the eastern (and more important) pyramid of Complex O. The placement of the western, four-stairway pyramid of Complex O also may have been the result of conscious planning, since it lies exactly true north as seen from Temple V. This pyramid has not yet been excavated, although it is generally supposed that it either immediately preceded or succeeded Temple IV. In this case, Temple V (Fig. 5) may have been constructed in the aforementioned time of transition and would fit with the choice of the place of the Western pyramid in Complex O (in AD 731). There also exists an interesting connection between TV and TIV, as seen from the doorway of TI. From TI, lines to TIV and TV (solid in Fig. 4) lie at an almost-perfect right angle to one another (our 1975 measurement gives 89°57'), each line deviating clockwise from the cardinal directions by 14° (consult Table 2 for details).

Fig. 5: View from the doorway of TI to TV (left) and the South Acropolis (unexcavated mound at the right).

Temple V may have been visualized with 3D-43, its somewhat smaller counterpart to the north, in mind. The importance of this south-facing structure is underlined by the discovery of a tomb beneath it in 1985, with an extraordinary stone monument of a human figure in the full round (seated cross-legged, the head missing). The shoulders and the back of the torso of this life-sized statue are covered by a vertically-arranged hieroglyphic text, containing calendric information and Tikal emblem glyphs. Interpretations are pending but should bring forth useful data (Mayer 1988). A visual line between V and 3D-43 (dashed in Fig. 4), which passes over a distance of nearly 1200 meters, is parallel to the main axis of the North Acropolis. As traced from the corresponding pyramids of Complex O, this same orientation may have been employed to determine the location of the two pyramids of the later Complex P of AD 751, immediately west of and adjacent to 3D-43.

There are at least two other indications that suggest a dependency of the placement and orientation of Temple IV upon earlier structures.[7] First, the direction from the doorway of Temple I to the doorway of Temple IV (solid line in Fig. 4) corresponds to the position of sunset on 12-13 Aug, the back-calculated commencement date of the Maya long count. The doorway of Temple IV also aligns at 90° to the axis of the North Acropolis and faces the no-longer-extant but once most important pyramid 5D-33-1st.

Second, a line perpendicular to the front of Temple IV, taken from its doorway, leads to the small platform 5E-

25 (about 300 meters east of Temple I), which is situated in the center of a great platform base (Structures 5E-23/28, of dimensions 100 x 120 meters and over 10 meters high) (This line is dashed and dotted in Fig. 4). The platform is quite similar to the one that supports Temple IV and could have served as a platform for a similar pyramid. "This might have become the biggest temple of all" said Guillemin (1968: 33). Could this have been the project of Ruler C, who built the two Twin-Pyramid Complexes Q and R immediately to the north?

The location for Temple III, the last of the great temples of the Ruler A dynasty, must have been carefully selected, for we find at least four lines that start or end here that might relate it to earlier pyramids. Perhaps most significant of all is the precise east-west direction (269°54') coming from the doorway of Temple I and passing through the center of the doorway of Temple III (solid in Fig. 4). This distance, nearly 320 meters, also forms the baseline of an isosceles triangle with its point of departure at the doorway of Structure 5D-104, a high temple-pyramid inside of the South Acropolis (Hartung 1977, p. 113).[8] Note also in Fig. 4 that this equinoctial alignment is parallel to the line between the center of Complex N and Structure 5D-33.

From the doorway of Temple III, a line (solid in Fig. 4) points exactly south to the doorway of Temple 5D-90, a building on the south side of the Court of the Seven Temples. This distance is about the same as that from Temple III to Temple I, and it represents another isosceles triangle with the point of departure at the doorway of Temple III (the two equal sides lie in the cardinal directions and form a right angle).[9] Finally, the doorway of Temple III can be related to the central doorway of Structure 5D-71, a building on the south side of the Great Plaza that confronts Str. 5D-33 on its north side, and which lies on axis with the North Acropolis. The direction to Temple III (dashed in Fig. 4) lies at 90° to the latter (the front of 5D-33 seems to have the same direction).

If all four of the lines we have suggested are considered as valid and deliberate constructions, then Temple III emerges as an exceptional example of the selection of a "special" place. The particular problem would have been to locate it in a center already built-up, rather late in Tikal's history, assuming the erection of Stela 24 in front of Temple III (9.19.0.0.0 = AD 810) refers to the inauguration of that pyramid.

Up to now, this has been the extent of studies of planning, arrangement and the orientation of buildings at Tikal. Whether the lines we posited a decade-and-a-half ago were deliberate or coincidental was in no real sense testable at the time and owing to the absence of other support evidence for the geometrical/astronomical hypothesis the situation for a long time has remained inconclusive. Today we believe Guillemin's (1968: 1-5) remark that "orientation would be a more religious concept than an urbanistic form" does not apply to the layout of Tikal. No longer is one justified in implying that mathematical and geometrical exactitude were never a consideration and even if considered, were always separated from religion in the Maya mentality. Just as in the codices, ritual and scientific views act together; they reinforce one another.

M. Miller's 1985 novel arguments about the idea of establishing order and form in an urban plan based upon principles of family relationships in a dynastic sequence may be just as tenuous as those arising from the method whereby one looks for purely geometrical/astronomical relationships; however, when the results of each independent method are compared, it is possible to find a number of points of convergence. Accordingly, the most salient fact to emerge from M. Miller's and our own study of the planning and arrangement of the monumental structures at Tikal is the central role of TI and its ancestral ties to 5D-33. Though by no means the largest structure at Tikal, TI nonetheless emerges as pivotal in both the genealogy and in the geometrical/astronomical analyses. The new order expressed by the axis TI - II and later TI - III and TI-IV runs counter to the north-south oriented Early Classic complex on the North Acropolis. TI is the center of the right angle TIV-I-V, which may imply that TV was placed exactly where we find it, not only to give an expansive view but also to tie it directly to the genealogical line IV-I. TI is also the pivot of the triangle TIII - TI - 5D-104. (At the present time we can say nothing more about complex 5D-104 genealogically.) And finally, it is the principal end of the west-east TIII-TI axis.

These links among specific Tikal structures are consistent with M. Miller's (1985: 13) conclusion that "none of the other temples was allowed to obstruct the view of the front elevation of TI." Despite the great number of constructions, not one interferes with the most important structure of all. "All others only revere and imitate TI, to whom all others look." We could go further by suggesting that in the spirit of the concreteness of dynastic imagery characteristic of the Ruler A dynasty, later rulers sought very definite ways to tie their structures to TI by placing them along significant and prominent visual and astronomical lines.[10]

Finally, it is important to note that archaeologists now have begun to take quite seriously the "new data and interpretations bearing on the nature of calendrical and cosmological rites in so far as they may influence site planning that have emerged in recent years" (Ashmore 1986). Thus, Ashmore (n.d.) has outlined the principal components of cosmological patterns including the use of ballcourts and causeways. At Tikal, the cardinal directions find an important expression in the twin-pyramid group as depicting "the Maya cosmos in miniature" (Ashmore n.d.: 4). Ashmore (n.d.: 6) also acknowledges that "Ah Cacau's constructions clearly emphasized the microcosmic layout of both the Great Plaza complex and the twin-pyramid groups".

In sum, we have tried to demonstrate that quite recently the data concerning persons related to specific buildings at Tikal have provided several strong points of reinforcement of earlier ideas we had posited solely on the basis of apparent geometrical and astronomical relationships among these buildings. Because we practice our disciplines in a world in which the written historical record takes precedence on the ladder of evidence, we believe that it would be valuable to continue to re-examine, in the way we have exemplified in this paper, the archaeoastronomical arguments already proposed at other Maya sites.

Acknowledgments:

We are grateful to Clemency Coggins and Chris Jones for their helpful advice and comments in the preparation of this paper.

TABLE 2.
Summary of Possible Astronomically-Related Alignments at Tikal

Alignment	Azimuth	Remarks
IV - I	104°	The sunrise dates are 14 Feb & 29 Oct. Whether these still might have any local agricultural significance is not known. They _do_ divide the year into 257/108 day periods, close to 260/105.
III (Facade)	108°16'	Faces generally east, but does not look directly at TI; also faces sunrise on _anti_-zenith passage. The same deviation of 18°16' S of E taken to the north (i.e. 18°16' N of E) goes directly to 5D-33. This is the direction of sunrise on the day it passes the zenith. Note (Aveni & Hartung (1986), Table 3) the special relation that obtains between the solstices and the zenith passage dates in the Petén area.
III - I	89°54'	Equinox sunrise.
I (Facade)	280°35'	Faces west. Solar dates corresponding to this direction are 18 Apr and 25 Aug. Twenty days before first solar zenith passage would correspond to an azimuth of 281°44'. If one looked over the top of TII (see Fig. 2) there would be a very slight adjustment for elevation.

Note: The azimuths quoted in arc-minutes were obtained with surveyor's transit and sun-fix at Tikal. These are accurate to 5 arc minutes.

NOTES

1. The most up-to-date table of the chronology of rulers at Tikal is given by Jones (1982: Table 6), a version of which appears in A. Miller (1986). Jones (p.c. 1/15/88) adds the following comments:
   (a) Under Stormy Sky, Jones disagrees with the assignment of Bu. 48 to Curl Nose.
   (b) Under Ruler B, St. 21 of TVI is the inaugural monument of this ruler.
   (c) Under Ruler C, the man depicted on St. 24 in front of TIII is Ruler C.

2. Though there is no wide agreement, some scholars reason that the Late Classic temples of Tikal were part of a master plan dedicated to the Ruler A dynasty, with rulership now being expressed in the form of concrete visual images rather than the more abstract images of the cyclicity of time employed in the Classic (see e.g. A. Miller 1986: 60).

3. Berlin's (1967) critique of the dismantling of 5D-33-1st is based strictly on archaeological grounds. See the replies by Rainey et al (1967) and Thompson (1967). These works deviated attention from the early observations of Maler and Tozzer by concluding that "the loss to mankind is not ... serious."

4. In contrast, M. Miller (1985) proposes that the lintel of TI (as well as that of TIV) depicts the dead ruler in an underworld confrontation. Jones' interpretation, that the ruler, very much alive, sits upon his throne on a wooden litter in celebration of a war victory, is based upon dates carved on the lintels that fall well within the lifetime of the ruler in each case.

5. That the doorways of TI and TV may have served as a departure point for a pair of lines that lead northward to the Twin Pyramids of Complex O was first suggested by Hartung (1977). Extended northward, these lines lead to complex P, which is similarly arranged. While these lines are not perfectly north-south, they do have the same direction as the principal axis of the North Acropolis. These complexes have long been regarded as symbolic of the quadripartite expression of the Maya cosmos (Jones 1977; Coggins n.d. and 1980). We measured the axes of Complexes Q and R and found them to lie in directions 2°10' and 5°28' E of N, respectively; i.e. quite close to the true N-S direction. Additional observations on possible geometrical relations at Tikal were contributed in an unpublished paper by Sanford (details reported by Hartung 1977: 117-118).

6. Coggins (p.c. 8/23/88), citing textual support in the TIV inscriptions, believes Ruler A died close to 9.15.2.0.0 and that TIV commemorates one lahuntun since his death. She suggests that Ruler B built TIV and that his mother is buried inside it.

7. Miller places Ruler B in Temple IV (but see Note 1); his pyramid is said to have been situated to keep the buildings of his parents in sight. Thus, TIV seems to be "in dialogue" with TI, the smaller TII deferring to them (indeed, his mother's back is literally turned to him). Thus, TIV gives reference at a distance from TI, even though it is larger. It "addresses" and it "does not challenge." Ruler C, to whom Miller assigns TIII, lies under the "sheltering shadow" of his father's TIV and its inscriptions make many references to his father. But the view from TIII also encompasses TI and TII; again, they are the primary referents. Thus, TIII indicates the family relationship to both sets of forebears; it lies in the "bosom of the dynasty." Given the placement of these structures, M. Miller goes on (p. 13) to speculate about the ruler to whom the eccentrically located TV is dedicated. She assigns Ruler AB (first son of Ruler A), in one of the date gaps between A and B, to this structure.

8. Today the South Acropolis is covered with high trees (Hartung 1977: 117, Fig. 9.3), but in former times, Str. 5D-104 would have been a very conspicuous building, as Fig. 3 shows.

9. Might the different orientation of Temple III (in comparison to the other great temples) have been chosen for a better visual relation with respect to the isosceles triangle?

10. Though the architectural tradition is quite different, the situation is reminiscent of the tomb of Pacal and the Temple of the Inscriptions at Palenque. Pacal died in AD 683 and surely had constructed his temple, while Ruler A had his inauguration in AD 682 and might have built 5D-33-1st, though not yet his tomb and the temple on top of it. In both cases, the doorways (the central one at Palenque) were the reference points, but at Palenque the tomb lies exactly below, while at Tikal it is north of the central axes. It would be rather unorthodox to obstruct the view of the front elevation of TI, and so it perhaps became more correct to tie one's structure to TI by placing it along prominent visual and astronomical lines.

REFERENCES

Ashmore, W. 1986 "Petén Cosmology in the Maya Southeast: An Analysis of Architecture and Settlement Patterns at Classic Quirigua", The Southeast Maya Periphery, ed. P. Urban and E. Schortmann, University of Texas Press, Austin, pp. 35-49.

Ashmore, W. n.d. "Deciphering Maya Architectural Plans", Paper presented at symposium on "New Theories of the Ancient Maya", University of Pennsylvania Museum, Philadelphia, 1987.

Aveni, A. & H. Hartung 1986 "Maya City Planning and the Calendar", Trans. Am. Phil. Soc. 76 (Part 7): 1-81.

Berlin, H. 1967 "The Destruction of Structure 5D-33-1st at Tikal", Am. Antq. 32: 241-2.

Carr, R. & J. Hazard 1961 "Map of the Ruins of Tikal, El Petén, Guatemala", Tikal Reports, no. 11, University of Pennsylvania Museum, Philadelphia.

Coe, W. 1967 "Tikal: a Handbook of the Ancient Maya Ruins", University of Pennsylvania Museum, Philadelphia.

Coggins, C. n.d. "Painting & Drawing Styles at Tikal: An Historical Iconographic Reconstruction" (Diss.) University of Michigan Microfilms, Ann Arbor, 1975.

Coggins, C. 1980 "The Shape of Time: Some Political Implications of a Four Part Figure", Am Antq 45: 727-739.

Guillemin, E. 1968 "Tikal: Development and Function of the Tikal Ceremonial Center", Ethnos 33: 1-35.

Hartung, H. 1975 "A Scheme of Probable Astronomical Projections in Mesoamerican Architecture", in Archaeoastronomy in Pre-Columbian America, ed. A.F. Aveni, University of Texas Press, Austin pp. 191-204.

Hartung, H. 1977 "Ancient Maya Architecture and Planning: Possibilities and Limitations for Astronomical Studies", in Native American Astronomy, ed. A.F. Aveni, University of Texas Press, Austin, pp. 111-129.

Jones, C. 1977 "Inauguration Dates of Three Late Classic Rulers of Tikal, Guatamela", Am Antq, 42: 28-60.

Jones, C. and L. Satterthwaite 1982 "The Monuments and Inscriptions of Tikal: The Carved Monuments", Tikal Report No. 33, University of Pennsylvania Museum, Philadelphia.

Mayer, K.H. 1988 "Tikal: A Unique Maya Sculpture on Exhibit", Mexikon 10, No. 3: 49-50.

Miller, A. 1986 "Maya Rulers of Time", University of Pennsylvania Museum, Philadelphia.

Miller, M. 1985 "Tikal, Guatemala: A Rationale for the Placement of the Funerary Pyramids", Expedition, 27, No. 3: 6-15.

Rainey, F., A. Kidder II, L. Satterthwaite, & W. Coe 1967 "Reply to Berlin", Am Antq, 32: 242-4.

Thompson, J.E.S. 1967 "A Third Party Comment", Am Antq, 32: 244.

Tozzer, A. 1911 "Preliminary Study of the Ruins of Tikal, Guatemala", Mem. Peabody Museum, Harvard 5(2), Cambridge, Massachusetts.

THE OBSERVATORY AT DZIBILCHALTUN

Clemency Chase Coggins and R. David Drucker

Introduction

At Dzibilchaltun two distant groups of structures that are connected by a ceremonial roadway are the halves of a single observatory. The eastern half of this macroassemblage has been excavated and reconstructed (Andrews IV and Andrews V 1980:82-148), its stucco decoration analyzed and its observatory function proposed (Coggins 1983:7); in this paper Coggins further proposes that the architectural form and serpent iconography of the eastern temple which combine Teotihuacan with Maya symbolism also encode observational and calendric data from the solar year, the Venus cycle, and the Maya Long Count- and that these were synchronized for a single day in Maya history: the completion of katun thirteen, 9.13.0.0.0 8 Ahau 8 Uo, or the Gregorian date March 16, A.D. 692. (1)

Dzibilchaltun is located in the dry northwestern corner of the Yucatan Peninsula 23 km from the Gulf of Mexico at 21 deg. 5 '45''North latitude. It is the first important site encountered traveling inland from the coast to the north and west (Kurjack 1979:13-15). Komchen, a site 6 km northwest of the center of Dzibilchaltun, was chosen for settlement in Middle Formative times (Ca. 700-400 B.C.) by people who were in contact, probably by sea, with the Olmec of La Venta to the southwest on the Gulf of Tabasco (Andrews V 1986:34-42), while greater Dzibilchaltun was populated until the Spanish Conquest, and may always have played some role in coastal trade.

In the seventh century more than a millenium after the founding of the Formative settlement a small structure (Str. 612) was built in the style of Teotihuacan 7 km southwest of the center of Dzibilchaltun. This cardinally-oriented single room building had a talud-tablero on its western, principal, facade (Andrews IV and Andrews V 1980:68-80). The talud-tablero is the hallmark of Teotihuacan architecture and no other such architecture is known from the northern Maya lowlands. About two centuries after this a massive west-facing structure (Str. 38) was built with "open-ended" talud-tablero terraces that suggest central Mexican or Cholulan inspiration (Andrews IV and Andrews V 1980:162). This unMaya construction, located at the central plaza of the site, was contemporary with the adoption of a widespread central Mexican 17.2 degrees east of north building orientation at Dzibilchaltun (Andrews V 1979).

The Observatory Assemblage

Sometime in the seventh century the ritual concourse of Dzibilchaltun was laid out with the Cenote Xlacah as its focus (Figure 1). The site's main axis is a sacbe, or ceremonial roadway, that extends east-west for 2,263 m on a line that is just 3.5 degrees south of east, or 93 deg. 30' azimuth. (2) A radially symmetrical temple (Str.66) at the west end of this sacbe had a single room, four doorways and four stairways, with two range buildings located to its east on either side of the sacbe axis. These three buildings are enclosed by a wall except on the east where the sacbe leads 200 m to a radial platform (Str. 63) with the plain Stela 21.

Over 2.5 km east at the opposite end of the axial sacbe another symmetrically radial temple (Str. 1-sub) mirrors the western one (Figures 1-4); it has four unusually wide doorways (235 cm wide) and four stairways, and is also enclosed by a wall, but this temple has a single range building to its east, and three pairs of such single-room buildings to its west, while a radial platform (Str. 12) with the plain Stela 3 is located 135m west of the group along the sacbe. Although they differ in details the east and west groups are twins, complementary halves of a single, entirely contemporaneous assemblage. In the Coggins report on the architectural decoration of the eastern buildings it was postulated that these widely separated groups were part of an equinoctial assemblage and analogous to the twin pyramid complexes of Tikal, and that like them its principal purpose was the celebration of the completion of a katun, and specifically of the katun ending at 9.13.0.0.0, in A.D. 692 (1983:52-54). Since no stelae or informative inscriptions date from this period at Dzibilchaltun these hypotheses were based entirely on the stylistic and iconographic traits of the architecture and its stucco decoration, and of associated ceramics.

Form and Decoration of the Eastern Temple.

The form of the eastern temple (Str. 1-sub), with its four-part radial symmetry is the three-dimensional equivalent of Mesoamerican signs that denote the completion of cycles: solar, planetary and calendric (Figures 3, 5; Coggins 1980, 1983:36-58). It is also the equivalent of the quincunx, or five-point sign that may denote Venus (Figure 5c). The upper facade of the temple had a stucco frieze with eight projecting heads located at the corners and at the center of each side all joined by two crisscrossing undulating serpentine forms that are studded with perforated discoidal elements (Figure 6). The heads have a curl at the corner of the upper jaw, and no lower jaw, although a tongue or beard-like element projects downward. They once had protruding noses. Their eyes are square with cross-eyed pupils, and each had a waterlily knotted on its forehead. In the Coggins analysis these were called "Chak" heads, or the head of God B which has a long nose and is usually associated with agriculture and the solar year. On the facade these serpentine forms swim in the sea and it was proposed that this represented the waters of the underworld from which the sun would rise directly to the east of this equinoctially-oriented temple (1983:7, 26, 27).

Fragments of stucco from the "tower" of Str. 1-sub and from other buildings in this group indicate such disc-studded serpentine forms with waterlilies and Cauac signs were once part of their (long gone) friezes as well, and that the central east-facing Str. 7 may have had a feathered serpent on its upper facade.

Teotihuacan Antecedents

While evidence of central Mexican contact at Dzibilchaltun is embedded in the local Maya cultural matrix certain of these traits will be identified here as background for an understanding of the observatory assemblage at the heart of the site. For instance, while very rare in the Maya Lowlands feathered serpents were commonly represented at Teotihuacan. The Temple of Quetzalcoatl in the Ciudadela at Teotihuacan is decorated with disc-studded tableros that contain feathered serpents, with different front and rear heads, that swim in a watery environment while single-headed feathered serpents swim along the taludes below (Figure 7). These feathered serpents probably represent Quetzalcoatl, the Venus creator deity who swam in the primordial sea; the differentiated heads of these Venus serpents correspond to the Morning (front head) and Evening (rear head) Star aspects of the planet and thus together denote Venus cycles (Coggins 1986b, 1987c,d).

The Temple of Quetzalcoatl is a west-facing eastern pyramid located within a walled precinct that is matched by another precinct on the western side of the main axis of Teotihuacan. Coggins has suggested that this serpentine eastern pyramid of Quetzalcoatl and its matching precinct served as a conceptual model for the east-west Dzibilchaltun observatory assemblage which was built five hundred years later (1983:42, 43) - a model both in orientation and in symbolism since the eastern Dzibilchaltun temple is similarly decorated with Venus serpents that swim in the (primordial?) sea. It is also located within a walled precinct and it is matched by a mirroring western precinct - although if there was ever a temple in the western precinct at Teotihuacan no evidence of it remains.

Probably equally important in a search for antecedents is the scale of the east-west sacbe at Dzibilchaltun which is two and one quarter kilometers long. The only other centrally located ceremonial roadway in all of Mesoamerica that its ambition and scale can be compared is the north-south Street of the Dead at Teotihuacan. Measured from the Pyramid of the Moon to the southern edge of the Ciudadela this is a hundred meters shorter than the Dzibilchaltun sacbe, although the Street of the Dead continues as far south again within the confines of the city (Millon 1973:42, fig. 6b). The orientations are different but the idea of laying out a road of such length as part of the preliminary planning of a sacred place is known in only these two instances.

It is the idea of a vast ceremonial axis with mirrored precincts dedicated to Venus serpents that is found at both Teotihuacan and Dzibilchaltun. Another important central Mexican concept that Coggins has postulated was integrated into Maya calendric celebration in the Early Classic was the fundamental organization of time into periods of 260, or of thirteen times twenty, units. This was the structure of the pan-Mesoamerican ritual calendar and it was given new emphasis in Maya Long Count celebrations through the influence of Teotihuacan emigrants at or near Tikal in the fourth century A.D. where it was expressed in a cycle of thirteen katuns which comprised thirteen twenty tun periods for a cycle of 260 tuns (Coggins 1979,1980). These new practises were numerologically evident in the particular symbolic significance of the numbers 13, 20 and 260 which later played key roles in the iconographic program of Str.1-sub at Dzibilchaltun - built to commemorate the completion of the thirteenth katun. In this thirteen katun cycle 260 tuns had been completed since the critical baktun completion date 9.0.0.0.0 8 Ahau 13 Ceh (Dec.11 435). (3)

The Iconography of the Eastern Temple

<u>The Long Count Dedicatory Date.</u> There are no contemporary inscriptions at Dzibilchaltun but the commemorative date 9.13.0.0.0 8 Ahau 8 Uo was encoded in Mayan iconography and language in the stucco decoration of Str. 1-sub for all to read (Figure 6). First, there are the serpents which Coggins has described as Chak-like water-lily serpents swimming in the waters of the underworld (1983:26,27,44-52); these have more recently been identified by Karl Taube as "an avian water serpent - a creature occurring as the personified tun glyph and as a head variant of the number thirteen" (1986:66,67; and see Hellmuth 1987:160-163; Thompson 1960:88, 136, 144). (4) In other contexts these serpents, their sinuous bodies lined with precious disc signs, connote the number thirteen while they are at the same time the personification of the tun, or 360 day year (Figure 8a,b). Each of the heads on the Dzibilchaltun facade once had a knotted water-lily on its forehead. (5) Such knotted forehead devices (in this case a water-lily which serves to identify the number thirteen, Figure 8a) may be read phonetically as the Yucatec Mayan syllable <u>k'a</u> from the verb <u>k'ax</u>, to tie or knot (Coggins 1987b, 1988b) and in association with the overlapping tun connotation these serpents could be read as <u>k'atun</u>, (6) <u>and</u> as "thirteen katuns", or the katun ending commemorated by the

construction of the temple.

The Calendar Round date 8 Ahau 8 Uo could probably also be read on this facade. There were eight serpent heads and the eight displayed the characteristics of the day and month signs. Unlike most such long-nose serpents the eight had the large square eyes with inward-focussed pupils of the sun god, the supreme lord, or Ahau - thus 8 Ahau may have been expressed, while Ahau was also the day that denoted the completion of the cycle. The serpents also had crossed bands on an element that projected downward in place of a lower jaw, and cross-hatched spots that were affixed to the sides of their eyes. The cross-hatched spots could be read, in Yucatecan, as ek' for black, and the crossed bands as k'at (Fox and Justeson 1984:48). Together these two elements combine to form the glyph for the month Uo (Figure 8c). In sum, the form of this temple expressed the completion of a cycle while the stucco decoration stated, both logographically and phonetically that the date of completion as thirteen katuns, 8 Ahau 8 Uo.

Solar and Venus Symbolism. The iconography of the facade of Str. 1-sub suggests other cycles besides those of the Maya Long Count. The long-nosed serpent itself may represent Chak, or God B of the codices, who presides over the 365 day solar and agricultural year as well as over a 260 day cycle that is synchronized with Venus (Bricker, this volume). The masks on this structure are also the earliest extant examples of the later common Chak facade masks of Yucatan and Campeche. At least four such masks crowned the four doorways of the circular Caracol at Chichen Itza, eight masks decorated the corners of Temple 22 at Copan, and Chak masks were arranged in groups of five on the upper facade of the Governor's Palace at Uxmal - these are the only three Maya structures that have been demonstrated to be Venus observatories (Aveni, Gibbs and Hartung 1975;, Closs,Aveni and Crowley 1984; Aveni 1980, pp. 273-277). (7) The presence of Chak masks need not designate an observatory but Maya observatories may all have Chak masks.

At Dzibilchaltun the serpents on the upper facade of Str. 1-sub are explicitly identified with Quetzalcoatl, Venus, and the sun, and they might have been read phonetically as kukulkan. In Yucatec Mayan the word for wavy or undulating is kukul while the word for serpent is kan. Kukulkan, undulating serpent, is a pun for K'uk'ulkan the Maya name for Quetzalcoatl. Both names mean feathered serpent, or bird-serpent, whether in Mayan or in Nahuatl.

If we look into some of the other Maya associations for the number eight a combined Venus/solar reading for this temple becomes even more obvious. Eight solar cycles equal five Venus cycles and the temple alludes to this formula in its eight serpent program, in its combined quadrilateral and quincunx form, and in its equinoctial and, as we shall see, Venus-viewing functions. The walled eastern group can also be viewed as containing eight structures if the doubled western buildings are counted separately, and as having five structures if they are counted jointly (Figure 3). Eight is also the median, and canonical, number of days that Venus disappears at inferior conjunction before its heliacal rise as Morning Star, as the Maya knew well. And for the Maya the eighth day is Lamat - "the day of the Venus God" (Thompson 1960:77).

<u>The Venus Serpent.</u> Tikal. The serpents that decorate Str. 1-sub are found in other Late Classic Maya contexts that confirm their Venus identity. One of these is an incised bone from the tomb of <u>Ah Kakaw</u> at Tikal. This fragmentary bone depicts a Maya canoe with five anthropomorphic passengers (Figure 9). The boat is apparently paddled, in the stern, by a figure wearing a crocodile headdress. In front of him a toad faces forward while a Maya lord, a monkey(?) and a mammal seated in the bow all face backward (Coggins 1975:480-483). The prow of the canoe is carved into a serpent head which has a a curl at the corner of its upper jaw and a bearded crossed bands emblem in place of a lower jaw. The beard streams forward as if the canoe were being paddled backward. The canoe has Lamat, or Venus, glyphs along its side and the head at the bow belongs to the Venus serpent.

This craft is further identified as a Venus boat by the presence, at the beginning of an inscription immediately above, of the "Venus over the earth" glyph which, with a repeated Calendar Round date (6 Akbal 16 Zac), is also found on other bones with canoe scenes from this same burial (Coggins 1975:469-479). It has been suggested that this glyph indicates the the date of warfare and the sacrifice of prisoners - possibly occuring at elongations or first appearances of Venus as Evening Star (Lounsbury 1982). Of the two possible Long Count dates for this inscription, 9.11.19.4.3, September 26, 671 and 9.14.11.17.3, September 14, 723, the earlier date is 18 days before Venus' Last Appearance as Morning Star. The later one is about 40 days before Venus' heliacal set as Evening Star, but not an elongation point; either date might be correct within the context of the burial. Whatever their significance the Venus connotations of this boat are clear in the Lamat glyphs, the "Venus over the earth" glyph, and the crossed bands "lower jaw" emblem which defines this aquatic serpent.

Copan. A second relevant example of this Venus serpent is found on Stela N at Copan. (Figure 10) Versions of its head form the headdress of the figure on the south face of the stela - a figure that is almost identical to the figure on the north. The two represent the Venus twins (the Morning and Evening Stars) on a stela that is dedicated to the Venus day One Ahau (Coggins 1988:99,100,107; 1986a). The twins wear Venus "serpent" headdresses with curls at the corner of the upper jaw, and with water lilies knotted on the forehead, like the Dzibilchaltun facade serpents. On the south figure smaller versions of this same head, also with the knotted water lilies and fish, and jaw curls, have crossed bands emblems in place of lower jaws as do the Tikal bone and the Dzibilchaltun serpent heads.

However, just as important as the shared Venus significance of this iconography is the calendric relationship between Copan Stela N and Str. 1-sub at Dzibilchaltun. They are dedicated to period endings 3 1/2 katuns apart: Str. 1-sub to 9.13.0.0.0 8 Ahau 8 Uo, March 16, 692; Stela N to 9.16.10.0.0 1 Ahau 3 Zip, March 15, 761. The two monuments are distinguished by the mechanisms they devised to commemorate both the Long Count date and the Spring Equinox. At Dzibilchaltun both dates (March 16 and March 20) were marked architecturally (Figure 11). On Stela N the Initial Series inscription records the equinox as well as the expected lahuntun ending which should read 9.16.10.0.0. 1 Ahau 3 Zip (March 15,761). The inscription actually reads 1 Ahau 8 Zip, five days later, or March 20 (the equinox).

A third example of this type of commemoration is the Castillo at Chichen Itza which, 3 1/2 katuns later, celebrated the completion of the baktun (10.0.0.0.0 7 Ahau 18 Zip) on March 13, 830 (Coggins 1983:55-57; 1986a, 1988b). The Castillo is almost as well known as an equinox marker (Rivard 1969) as it is for its feathered serpent iconography which is usually read as Kukulkan, or Venus, symbolism. Unlike at Dzibilchaltun and Copan we do not know how the seven day interval between March 13 and 20 was registered in the Castillo, but these three March period endings, separated by 3 1/2 katuns, share Long Count, equinoctial, and Venus symbolism. If Str. 1-sub was not the first of its kind the katun ending 9.6.0.0.0 9 Ahau 3 Uayeb must have been commemorated somewhere in analogous fashion, since it fell on March 20.

The association between the Spring equinox and Venus may have had to do with the even balancing of the year between halves that were metaphorically analogous to the Morning and Evening Star phases of Venus, whereas the close association with the period endings was cyclic (during the Classic period). It is possible to read in its form and decoration the exact Long Count date for which Str. 1-sub was built as well as to deduce that the temple was probably involved in Venus and solar related ritual. This information alone might be enough to date it - but when we consider some of the astronomical phenomena that the Dzibilchaltun Maya could have observed in this observatory there can be no question of the specific dedicatory date.

Astronomical Observation and the Eastern Temple

Solar. In previous work on Str. 1-sub Coggins suggested this eastern temple was used as a sunrise foresight when viewed from Str. 66 on the west (1983:7). This was especially likely because Str. 1-sub has four doorways aligned on the east-west axis with visibility through to the eastern horizon, and the axial doorways are flanked by windows, to the north and south, which also give views through to the far horizon (Figures 2-4,11). Few, if any, vaulted structures existed at Dzibilchaltun when the sacbe was laid out and the twin temples built (Andrews IV and Andrews V 1980:123-124), and thus no big buildings or natural features obstructed the views to the flat eastern or western horizons. The doubled assemblage was probably built to create calibrated horizons for the recording of astronomical events where there was no significant natural topography that might be used for this purpose.

When Hartung and Aveni measured the sight line from Str. 66 on the west to Str. 1-sub on the east they found this axis was skewed 3.5 degrees to the south of east. and they found the same deviation when they sighted through the Str. 1-sub doors from the stela 3 platform (Str. 12) just to the west of the eastern group. They felt confident such a consistent deviation was deliberate (Figures 2,3). From either vantage point the sun would have risen through the four axial doorways four days before the spring equinox, whereas it rose through the two aligned southern windows six days before, and through the two aligned northern windows two days before the equinox. On this same day the sun would probably also have risen above the center of the temple tower (Figure 11). (8)

The windows apparently served as anticipatory, or predictive, markers for the sun's appearance in the doorway at a time of year when it was moving most rapidly, and was thus hardest to "catch". (9) In the fall this sequence of events would have been reversed; as the sun approached the tower from the north it would have been visible in the windows and doorways six, four, and two days <u>after</u> the fall equinox. While the Aveni and Hartung measurements confirmed the equinoctial functions of this building they did not explain the 3.5 degree skew, beyond its predictive character. Their measurements did, however, reveal other possible observational uses for this eastern group (Figure 3). E. Wyllys Andrews V informs Coggins that as Strs. 6-7 are reconstructed only the very top of the Str. 1-sub doorways, and not the windows, would have been visible from the Stela 3 platform (Str. 12). Thus solar viewing eastward <u>Through the doorways</u> was probably only possible from Str. 66, on March 16, four days before the equinox.

Much of the eastward horizon viewing probably took place on the Stela 3 platform. (Figure 3; Table 1) After the equinox, March 20, the first zenith occurred on May 25, and the second, after an interval of 52 days, on July 17 (Chart 1). (10) The two zenith sunrises would have been visible (from Stela 3) to the north of the three symmetrical pairs of facade buildings above the center of the smaller Str. 10 which had been added to the group (Figures 3, 11). When it was built the architectural form of Str. 10 was most unusual since it consisted of a single room with four cardinally placed doorways. The western one was 250 cm wide, wider than the Str. 1-sub doorways; it was later narrowed and the other three were filled in (Andrews IV and Andrews V 1980:132). Str. 10 was also unlike other buildings in the group in having had glyphic sun-related ahau and kin elements in its stucco facade decoration (Coggins 1983:29-31). The corners of this building, the east-west doorways, and some parts of the roof decoration served as markers for zenith and summer solstice sunrises.

The 52 day interval between the equinoxes is significant because fifty two is one of the most numerologically important mesoamerican numbers (2 x 26, 4 x 13, and the number of years in a Calendar Round cycle); Milbrath reports that agricultural calendars in the Maya Madrid Codex are laid out in 52 day periods (1979:278-279). Observation of the two zenith passage dates, which vary according to latitude, were also particularly significant in ancient Mesoamerica and this 52 day interval may have been instrumental in determining the location of this observatory group (in Yucatan to the east, Motul, and less precisely Tizimin and Culuba, had this same interval between zenith passages). Copan, far to the southeast, has an interval twice as long (Aveni 1977:fig. 1.5).

On the south side of the group winter solstice may have been marked over a small platform that was added to the south end of the two southwestern buildings (Strs. 4-5), or possibly from Structure 5 itself over Structure 3, while the actual solstice sunrise might have been sighted from the later higher Str. 11, to the southwest, along the north wall of the strangely skewed Str. 2 at the eastern edge of the group.

Long Count commemoration. Observation and celebration of the predictable structure of the solar year was only one of the purposes of this group - and it was possibly the least important. It is the 3.5 degree deviation from east that gives the conclusive clue to the dedicatory date and paramount commemorative purpose of the eastern temple. The sun rises through the four aligned doorways four days before the vernal equinox, or on March 16 which in A.D. 692 was the day 8 Ahau 8 Uo, 9.13.0.0.0. that commemorated the completion of the thirteen katun cycle of 260 tuns since the beginning of baktun nine. Str. 1-sub was built to capture this particular sunrise, as was the whole macroassemblage, and this date probably signified the Classic period founding of Dzibilchaltun. This dedicatory date could have been celebrated in this way only until Str. 1-sub was superseded perhaps a century after its construction, although observations of the stations of the solar year did not become obsolete, and the exteriors of Structures 4-9 were never modified. (11)

Venus as Morning Star. The solar observations were keyed into another calendric observation that, in view of the serpentine decoration of the facade, must be viewed as having been as important as the katun commemoration. This involved tracking the Morning Star appearances of Venus which is so closely tied to the course of the sun. As noted above five Venus cycles equal eight solar cycles, and each Venus cycle includes a period as Morning Star and a period as Evening Star. In discussing these appearances Anthony Aveni recently illustrated his paper with charts that show the height reached each day by Venus and the configuration these positions make during each period of appearance (Aveni 1986); see Figure 12 for the five Morning Star phases of the Venus cycle that began in November 691. (12)

In the Dresden Codex, the best source for evidence of the Maya observation of Venus, the number of days assigned for the eastern (morning) and western (evening) appearances (236 and 250 respectively) are probably of numerological significance since they are not observationally correct. Aveni argues that each interval was actually understood to last 260 days since the mean period is 263 days. Two hundred sixty days is also the length of the ritual calendar, and 260 tuns the length of a katun cycle. The importance of this number was noted above, and Aveni plausibly proposes that it was known as the length of each Venus appearance as well.

The Five cycles. The striking Venus configurations on the charts (Figure 12) trace the distance from the sun that Venus attained at five day intervals during the course of its approximately 263 days of appearance as Morning Star: Venus would have been visible at only one of these points at a time so that while rise points might easily be marked along the horizon (real or man-made) relative elevation would be much more difficult to record.

During the course of its five cycles Venus traces five different patterns in the eastern morning sky, and five in the western evening sky. These have markedly different characters. The one we will consider the first of the five at Dzibilchaltun began with the heliacal rise of Venus late in November to the south of east and ended in August with rise to the north, thus including the vernal equinox, first and second zenith passages, and summer solstice in its period as Morning Star (Figures 12a,13; table 1; chart 2). The second cycle began early in July, 584 days (one Venus cycle) after the first, with heliacal rise to the north of east and ended with rise in February to the south, thus including second zenith passage and the autumnal equinox and winter solstice (Figure 12b) The third began with a February heliacal rise to the south of east and ended to the north in October; it inlcuded both equinoxes, both zeniths, and summer solstice (Figure 12c). The fourth began in September with heliacal rise just north of east and ended late in April not far from the same horizon point; it included both equinoxes and the winter solstice (Figure 12d). The fifth began in April with heliacal rise to the north of east and ended with December rise to the south; it included both zeniths, summer solstice, fall equinox, and possibly winter solstice (Figure 12e).

Each of these periods of eastern (and of western) appearance traces its own characteristic pattern in the sky, with elongations and retrograde motion creating loops that, along with each distinctive seasonal character, helps explain the differentiated personalities ascribed to the five periods in the illustrations of the Dresden Venus Table (Thompson 1972:ms. pp. 46-50). (13) The fact that over the five periods the heliacal rise shifts along the horizon from south of east to north of east, to south, to north, and finally to north again adds to the distinctiveness of these appearances and would have keyed them into the architecture of the eastern group at Dzibilchaltun where I believe they were tracked. (14)

If the tracks of the five Morning (and Evening) star phases were superimposed, on paper or in the minds of the observers, they would appear to criss-cross, or interweave because four of the five consecutive paths alternate north to south and south to north. Coggins proposes that this was symbolized by the criss-crossing bodies of the Venus serpents on the facade of the temple (Figure 6), and also by the serpents' crossed bands emblems. The crossed-bands glyph, T552, is often part of the sky glyph, T561c, and is regularly part of "sky bands". It was probably read k'at, "crossing", in Yucatecan (Kelley, 1976:152), and refers to the apparently erratic path of planets, particularly of Venus. Such an interpretation would also explain the importance of the association of Venus on Str. 1-sub with the month Uo whose glyph, as noted above, could be read ek k'at in Yucatecan. This name may mean "black crossing", or it could mean "star crossing", and thus serve as a name for Venus.

The evidence for tracking Venus as Morning Star in this group ties the heliacal rise of Dzibilchaltun Venus cycle 1 to the group's March 16 dedicatory date and to the end of the katun cycle (Table 1; Figure 11). On March 16, 692, Venus rose at the same horizon point that it had 83 days before in its heliacal rise. But perhaps the most extraordinary aspects of this Morning Star period are the Venus cycle's critical Calendar Round dates. In Dzibilchaltun cycle 1 Venus appears to have followed an ideal pattern and this particular cycle might therefore have served as a canonical model (Chart 2).

Heliacal Set as Evening Star. The last appearance of Venus as Evening Star occurred on November 17, 691. (15) This was the numerologically significant Long Count date 9.12.19.12.0 5 Ahau 13 Kankin which is 13 completed days after the thirteenth mesomonth of the Calendar Round. It was also the beginning of the thirteenth uinal in the closing katun thirteen. Thirteen was also personified in the temple's Venus serpent decoration, as noted above. The glyph for the month, Kankin, in which the last appearance occurred, represented the dog that signified Venus as Evening Star (Closs 1979, p. 163); the day sign Ahau denoted Lord, while the day number, five, was the quintessential Venus number as Aveni has pointed out (1986).

Heliacal Rise. After this last appearance the Venus period of disappearance at inferior conjunction lasted exactly eight days as it is canonically described in the Dresden Codex. This is the median, but the period may be as short as zero and as long as twenty days (Aveni 1986). The heliacal rise of Venus in Dzibilchaltun cycle 1 was November 25, A.D. 691 in the Morning Star phase that was to include the March 16 completion of katun thirteen and the presumed dedication of the architectural assemblage. This Long Count date was 9.12.19.12.8 13 Lamat 1 Muan. Like the heliacal set date this one combines thirteen with a Venus name since Lamat is the day and glyph associated with Venus.

Intermediate Points. During this cycle as Morning Star Venus returned to its 108 degree heliacal rise position on the dedicatory day 8 Ahau 8 Uo, 9.13.0.0.0, after having risen almost as far south as the winter solstice rise point and then looped back toward the north. On both those days Venus rose over the center of the southern Structures 4-5 (Figures 3,11; Table 1). The equinox was four days later. On April 21 Venus rose through the doorway of Str. 1-sub. This was four days before the end of the 40 day (two uinal) period since March 16; (16) this is another example of the use of a four day anticipatory interval.

On May 17 Venus rose over the the northern Structures 8-9. This was four days before May 21 which was the heliacal rise of the Pleiades, visible through the axial doorways of Str. 10 (fig. 11). These two dates may have been of particular historical significance at Dzibilchaltun because May 17, first zenith passage at Teotihuacan, had coincided with the heliacal rise of the Pleiades when the Teotihuacan orientation was laid out - while at that time the Pleiades also set on the Teotihuacan orientation line (Aveni and Gibbs 1976:516). An additional four days later, May 25, was first zenith sunrise at Dzibilchaltun. Like the Pleiades the zenith sun rose on the axis of Str. 10. This relationship, memorialized in this bulding at A.D. 692, linked the Pleiades and the Zenith - perhaps in emulation of the Teotihuacan relationship. Str. 10's east doorway may have been closed off when the Pleiades were no longer visible there - a century or more later.

Last Appearance. On the day of its last appearance Venus rose at the same horizon point it had on May 25 (First Dzibilchaltun zenith passage). Last Appearance occurred 260 days after heliacal rise, on August 10, 692-9.13.0.7.7 12 Manik 15 Chen. This date completes the ideal 260 day Morning Star period - in the same period as the completion of the 260 tun katun cycle. Whereas the next day, 13 Lamat 16 Chen was August 11, "Maya Creation Day", or the solar anniversary of 4 Ahau 8 Cumku the date of the beginning of the Long Count in 3114 B.C., 9.13.0.0.0 baktuns earlier, at the end of the preceding 13 baktun cycle. That mythical earlier cycle was the ultimate thirteen date, one that was perhaps invoked by the <u>K'uk'ulkan</u> facade serpents if they were understood as swimming in the primordial seas of the creation (Coggins 1986).

"Maya Creation Day", August 11, is also a Teotihuacan orientation day (one of the two days on which the sun sets on the orientation axis as viewed from the Pyramid of the Sun), while it is also second zenith passage at Izapa, Kaminaljuyu and Copan (Drucker 1977). We suggest that sunset on August 11-12 may be the basis for the "17 degree family of orientations " (Aveni 1975:166-168) found throughout Mesoamerica. Sunset on August 11-12 may have determined the location of Temple IV, relative to Temple I, at Tikal (Aveni and Hartung, this volume), and further research may reveal the significance of this date for orientation purposes throughout Mesoamerica.

The following period of superior conjunction at Dzibilchaltun was 52 days long if Venus' first appearance as Evening Star was celebrated on October 2, 692, which was 9.13.0.10.0 13 Ahau 8 Ceh, another important thirteen date (although Venus may have appeared two days earlier). (17) In sum, this Venus cycle consisted of eight days at inferior conjunction, 260 days as Morning Star, 52 days at superior conjunction and 264 days as Evening Star. These may have been canonical Venus periods that were authorized in 692 by the construction of this Dzibilchaltun observatory. The Evening Star appearances of Venus were surely also tracked for this particular cycle but the ruined western group must be measured from its stela platform to understand how it may have worked.

<u>Observatories</u>

The eastern half of the Dzibilchaltun observatory was the larger group with a bigger pyramidal structure and it was located closer to the center of the site (Figure 1). The smaller western half, which was probably observed from both Str. 1-sub and from its associated stela platform, was also constructed differently in that there were only two range buildings between it and the platform. Thus it would have been possible to sight between the buildings from this platform and through Str. 66 if it had axial doorways like Str. 1-sub. While they differed in detail, the two ends of this macroassemblage were reciprocally related - like the opposing differentiated heads of the bicephalic serpent (Figure 7). The eastern "head" connoted the rising sun and Venus as Morning Star and the western "head" the setting sun and the Evening Star.

No other observatory like this one is known from any date. The preclassic equinox and solstice observatories at Tikal (Fialko 1988) and Uaxactun (Aveni and Hartung 1986b) also sighted eastward with the aid three small aligned structures, but from large radial pyramidal structures. As at Dzibilchaltun the solstices were sighted at the northern and southern edges of the eastern group and the equinox may have been viewed through an axial doorway at an early stage of their use, but the pyramidal structures in all of these groups eventually became so high that horizon viewing could not have involved intermediary buildings as foresights. Instead backsights were probably incoporated into the sighting structure as was later done with the windows of the Caracol (Aveni, Gibbs and Hartung 1975), or an interior wall was used as in the Temples of the Jaguars at Chichen Itza (Milbrath, this volume).

As katun commemorating complexes the twin pyramid complexes at Tikal are symbolically analogous to the Dzibilchaltun observatory since they include east-west mirroring radially symmetrical pyramids (Coggins 1983, pp. 40-41). If they were observatories the horizons might have been viewed from the two large pyramids in the same way as from the later enlarged pyramids at Tikal, Uaxactun, and Dzibilchaltun, or they could have served as foresights for each other. Early Maya observatories were oriented toward the east (Ruppert 1940), whereas the Pyramid of the Sun and the Temple of Quetzalcoatl at Teotihuacan, as well as the much later Caracol at Chichen Itza were oriented toward the west. The twin pyramid complexes at Tikal may have formalized dual observation, the observatory at Dzibilchaltun may have refined the idea, and at Chichen Itza, the Castillo, may have institutionalized it.

The Temple of Quetzalcoatl at Teotihuacan could have been used for observation of the eastern horizon from the radial platform to its west, as at Dzibilchaltun, or possibly from the pyramid itself and across the three small structures on the eastern wall of the compound, as Fialko proposes, (1988, pp. 19, 20). But when foresight structures are smaller than the viewing pyramid, as at Teotihuacan, in the later phases of the Tikal and Uaxactun groups, or at either end of the Dzibilchaltun assemblage, then it is likely that it was the nearby low axial radial platforms that served as the viewing sites. The stela platforms at Dzibilchaltun, located at some distance down the sacbe from the eastern and western groups were evidently viewing platforms for their associated temples. Each had a plain stela that may also have served some sighting purpose as well as functioning as zenith gnomons, and as commemorative of the 9.13.0.0 0 date for which the assemblage was built.

Conclusions

The observatory at Dzibilchaltun was built with a four day predictive capacity that served to synchronize the solar, Long Count and Venus dramas that were enacted there. The fact that the primary focus of the assemblage was a date, and not just an astronomical event, was probably true for the orientation of all Mesoamerican religious structures. Anniversaries such as "Maya Creation Day" (August 11-12) which had scriptural significance for all Maya as well as historic significance at Teotihuacan might determine orientations just as the completion of a major calendric period, in conjunction with a solar event like spring equinox, might combine with an ideal Venus cycle to determine the layout of a sacred place where the interval between zenith passages was exactly 52 days. This paper has sketched only some of the more obvious horizon events at Str. 1-sub where all the stars were players in the heavenly dramas for which the Maya had constructed the proscenium.

ACKNOWLEDGEMENTS

Much of the raw data for this paper was supplied by others to whom we are exceedingly grateful. Anthony Aveni and Horst Hartung made the measurements at Dzibilchaltun and offered suggestions for possible solar sight lines. Owen Gingerich and Barbara Welther of the Harvard-Smithsonian Center for Astrophysics generously created and provided the Venus cycle charts. Victoria Bricker, E. Wyllys Andrews V, and Horst Hartung also made valuable comments on the first draft.

NOTES

1.In this paper Gregorian dates are used with the Thompson 11.16.0.0.0 correlation and Julian day number 58423.

2. Azimuth measurements in this paper were planned and made by Anthony Aveni and Horst Hartung. They differ by 1 deg. from the Dzibilchaltun site map which was laid out in reference to this east-west sacbe at 4.5 deg. east of true north (Kurjack 1979, p. 4).

3. The completion of the ninth baktun at 9.0.0.0.0 8 Ahau 13 Ceh (A.D. 435, December 11), and the tenth at 10.0.0.0.0 7 Ahau 18 Zip (A.D. 830, March 13), were historically associated with Teotihuacan influenced Maya prophecy (Coggins in press b).

4. Karl Taube also points out that the personification of the number thirteen is depicted in a graffito on the floor of Str. 1-sub (1986:67). In its published illustration (Andrews IV and Andrews V 1980: fig. 110) this head is not clear because the figure is turned 90 deg. counter-clockwise.

5. Only six of the eight masks retained any stucco decoration, but these fragments suggest the eight masks were iconographically equivalent (Coggins 1983: 26,27).

6. <u>K'atun</u> signifies both the tying up of the tun cycle as well as denoting 20 tuns since <u>k'ax</u> means tie and <u>k'al</u> means 20 (Coggins 1987b).

7. Herbert Spinden noted the associations of Chak (the long-nose deity) with Venus, the sun (kin sign), the bicephalic monster, and the manikin scepter deity (God K) (1913:61-69; and see Kowalski 1987:187-202). The same cluster of associations may be made here since the first three are inherent in the symbolism of the Dzibilchaltun observatory assemblage. The manikin scepter is not evident and probably only later became important at Dzibilchaltun (Coggins in press, in preparation).

8. The original height of this reconstructed tower is unknown.

9. Real windows like these are otherwise unknown in Maya architecture (Andrews IV and Andrews V 1980:95).

10. Drucker calculated zenith dates from the <u>American Ephemeris and Nautical Almanac</u> (1968) using the solar declinations after leapday for the first year of four.

11. Str. 1-sub. was built with an experimental annular vault that may have collapsed and the building was covered by the much larger radially symmetrical Str. 1 (Coggins 1983, pp. 7,8,56).

12. Owen Gingerich provided Coggins with sets of configurations for Venus cycles in the years A.D. 689-698 and 721-730, while Anthony Aveni shared his A.D. 500-509 and 650-659 sets. These were all prepared by Barbara Welther and Owen Gingerich of the Harvard-Smithsonian Center for Astrophysics from a program devised by Barbara Welther, Owen Gingerich, and Beth Collea.
12.

13. These cycles are not fixed in the year, although the configurations change very little. Dzibilchaltun cycle 1 was about twelve days later in A.D. 651-652 than it was in 691-692 and it was ten days earlier in 723-724, while the rise points were a degree to the south in 651 and a degree north in 723 (109-107 degrees azimuth in 9 cycles, or 72 years).

14. This pattern of Venus rises to the north and to the south of east (and also of west) is illustrated in the wall paintings of the Upper Temple of the Jaguars at Chichen Itza where the east and west axes of the paintings are clearly designated by two figures that apparently signify the sun and Venus. At the same time Venus is shown rising in separate paintings just to the north and south of these axial portraits. Venus is shown at superior conjunction on the north wall of this temple room and at inferior conjunction on the south, so that the cycle of paintings corresponds to a complete Venus cycle of 584 days (Coggins 1984, pp.156-165).

15. Drucker consulted Tuckerman's tables (Tuckerman 1964), with corrections from Neugebauer (1938).

16. See Drucker 1977, Aveni 1977, and Aveni and Hartung 1986a for calendric uses of twenty and forty day intervals.

17. 9.13.0.10.0 13 Ahau 8 Ceh would have been desirable for its period ending and for its 13 and 8 coefficients.

1. Heliacal Rise of Venus, November 25, 691. Center Str. 4.

2. Winter Solstice, December 21, 691. Platform, S. Str. 4.

3.* Sunrise. March 14, S. windows, Str. 1-sub.

4.* Sunrise. March 16. Axial doorways, Str. 1-sub.
Venus Rise March 16. Center Str. 4.

5.* Sunrise. March 18. N. windows, tower, Str. 1-sub.

6. Spring Equinox, March 20. N. corner Str. 1-sub.

7. Pleiades Rise, May 21. Axial doorways Str. 10.

8. First Zenith Passage, May 25. Center Str. 10.
Venus Rise, May 25. Between Strs. 8-10 (see Aug 10).

9. Summer Solstice, June 21. N. corner Str. 10.
* Orion Epsilon rise, June 26. S. window Str. 1-sub.

10. Second Zenith Passage, July 17. Axial doorways Str.10.

11. Venus Last Appearance, August 10. Between Strs. 8-10. (see May 25).

12. Sunrise, Maya Creation Day, August 11. N. end Str. 8.

13. Fall Equinox, September 23. N. corner Str. 1-sub.

Table 1. Postulated observations of astronomical horizon events at the Str.1- sub group, 691-692 A.D. Starred (*) sightings visible from Str. 66, all others from Str.12.

BIBLIOGRAPHY

Andrews IV, E. Wyllys and Andrews V, E. Wyllys

1980 Excavations at Dzibilchaltun, Yucatan, Mexico. Middle American Research Institute Publication 48, Tulane University, New Orleans.

Andrews V, E. Wyllys

1979 Early Central Mexican architectural traits at Dzibilchaltun, Yucatan. Actes, 42nd International Congress of Americanists, VIII:237-249, Paris.

1986 Olmec jades from Chacsinkin, Yucatan and Maya ceramics from La Venta, Tabasco. In Research and Reflections in Archaeology and History, editor E.W. Andrews V., Middle American Research Institute Publication 57, pp. 11-47, Tulane University, New Orleans.

Aveni, Anthony F.

1975 Possible astronomical orientations in ancient Mesoamerica. In Archaeoastronomy in Pre-Columbian America. ed. A.F. Aveni, pp. 163-190. University of Texas Press, Austin.

1977 Concepts of positional astronomy employed in ancient mesoamerican architecture. In Native American Astronomy. ed. A.F. Aveni, pp. 3-19. University of Texas Press, Austin.

1980 Skywatchers of Ancient Mexico. University of Texas Press, Austin.

1986 The real Venus-Kukulcan in the Maya inscriptions and alignments. Sixth Mesa Redonda de Palenque. University of Oklahoma Press, Norman. In press.

Aveni, Anthony F., S. L. Gibbs, and H. Hartung

1975 The Caracol tower at Chichen Itza: an ancient astronomical observatory? Science, 188 (4192): 977-985.

Aveni, Anthony F., Sharon L. Gibbs

1976 On the orientation of ceremonial centers in Central Mexico. American Antiquity, 41:510-517.

Aveni, Anthony F., Horst Hartung

1986a Maya city planning and the calendar. Transactions: American Philosophical Society vol. 76, part 7:1-87.

1986b Uaxactun, Guatemala, Group E and similar assemblages: an archaeoastronomical reconsideration. In World Archaeoastronomy, editor, A. F. Aveni. Cambridge University Press. In press.

Closs, Michael P.

1979 Venus in the Maya world: glyphs, gods and associated astronomical phenomena. In Tercera Mesa Redonda de Palenque, vol. IV, editors M. G. Robertson and D. C. Jeffers. Herald, Monterey. pp. 147-165.

Closs, Michael P., A.F. Aveni, and B. Crowley

1984 The planet Venus and Temple 22, Copan. Indiana, 9:221-248.

Coggins, Clemency Chase

1975 Painting and Drawing Styles at Tikal: an Historical and Iconographical Reconstruction. Ph.D. dissertation, Harvard University. University Microfilms, no. 76-3783. Ann Arbor.

1979 A new order and the role of the calendar: some characteristics of the Middle Classic period. In Maya Archaeology and Ethnohistory, editor Norman Hammond. University of Texas Press, pp. 38-50.

1980 The shape of time: some political implications of a four-part figure. American Antiquity, 45:727-739

1983 The Stucco Decoration and Architectural Assemblage of Str. 1-sub, Dzibilchaltun, Yucatan, Mexico. Middle American Research Institute Publication 49, Tulane University, New Orleans.

1984 Catalogue.In Cenote of Sacrifice: Maya Treasures From the Sacred Well at Chichen Itza. editors C. C. Coggins and O.C. Shane III. University of Texas Press, Austin.

1986a New Sun at Chichen Itza.In World Archaeoastronomy. editor, A.F. Aveni. Cambridge University Press. In press.

1986b Reflections on Teotihuacan. Paper presented at the Society for American Archaeology meetings, New Orleans.

1987a New fire at Chichen Itza. Memorias del Primer Coloquio Internacional de Mayistas, (1985), Universidad Nacional Autonoma de Mexico, pp. 427-484.

1987b The names of Tikal. Primer Simposio Mundial sobre Epigrafia Maya, Asociacion Tikal, Guatemala. pp. 23-45.

1987c Quetzalcoatl, Teotihuacan, and the Maya. Paper presented at the Society for American Archaeology meetings, Toronto.

1987d Names of Quetzalcoatl at Teotihuacan and in Maya regions. Paper presented at the American Anthropological Association meetings, Chicago.

1988a On the historical significance of decorated ceramics at Copan and Quirigua and related Classic Maya sites. In The Southeast Classic Maya Zone, editors E.H. Boone and G.R. Willey. Dumbarton Oaks, Washington, D.C., pp. 95-123.

1988b The birth of the baktun at Tikal and Seibal. In Vision and Revision in Maya Studies, editors F. Clancy and P.D. Harrison. University of New Mexico Press. In press.

Drucker, R. David

1977 A solar orientation for Teotihuacan. Los Procesos de Cambio (en Mesoamerica y areas circunvecinas). XV Mesa Redonda, Sociedad Mexicana de Antropologia y Universidad de Guanajuato. vol. 2: 277-284.

Fialko, Vilma

1988 Mundo perdido, Tikal: un ejemplo de Complejos de conmemoracion astronomica. Mayab 4:13-21. Madrid.

Fox, James A. and John S. Justeson

1984 Polyvalence in Mayan hierogolyphic writing. In Phoneticism in Mayan Hieroglyphic Writing, editors J.S. Justeson and L. Campbell. Institute for Mesoamerican Studies, State University of New York Publication 9: 17-76. Albany.

Hellmuth, Nicholas

1987 The lily-pad headdress monster, Monster und Menschen in der Maya-Kunst. Graz. pp. 160-163.

Kelley, David H.

1976 Deciphering Maya Script., University of Texas Press, Austin.

Kowalski, Jeff Karl

1987 The House of the Governors: a Maya Palace at Uxmal, Yucatan, Mexico. University of Oklahoma Press, Norman.

Kurjack, Edward B.

1979 Introduction to the Map of the Ruins of Dzibilchaltun, Yucatan, Mexico. Middle American Research Institute Publication 47, Tulane University, New Orleans.

Lounsbury, Floyd G.

1982 Astronomical Knowledge and its uses at Bonampak, Mexico. In Archaeastronomy and the New World, editor, A.F.Aveni. Cambridge University Press, pp. 143-168.

Milbrath, Susan

1981 Astronomical imagery in the serpent sequence of the Madrid Codex. In Archaeastronomy in the Americas, editor R.A. Williamson. Balena Press, Anthropological Paper no. 22, Los Altos, CA.

Millon, Rene

1973 The Teotihuacan Map, Part I: Text. In Urbanization at Teotihuacan, Mexico, editor R. Millon, University of Texas Press, Austin.

Neugebauer, P.V.

1938 Tafeln zur Berechnung der Jahrlichen auf- und unter-gange de planeten. Astronomische Nachrichten 264(6331): 313-322. Kiel.

Ruppert, Karl

1940 A special assemblage of Maya structures. In The Maya and Their Neighbors. editors C. Hay et al. Appleton-Century, N.Y. pp. 222-231.

Spinden, Herbert

1913 A Study of Maya Art. Memoirs of the Peabody Museum vol. VI. Harvard University, Cambridge, MA.

Taube, Karl

1986 The Teotihuacan cave of origin. Res 12: 51-82.

Thompson, J. Eric S.

1960 An Introduction to Maya Hieroglyphic Writing, University of Oklahoma Press, Norman.

1972 A Commentary on the Dresden Codex. American Philosophical Society Memoirs vol. 93, Philadelphia.

Tuckerman, B.

1964 Planetary, Lunar, and Solar Positions, A.D. 2 - A.D. 1649 at 5 Day and 10 Day Intervals. American Phlosphical Society Memoirs vol. 59, Philadelphia.

U.S. Government printing Office

1968 American Ephemeris and nautical Almanac. Government Printing Office, Washington, D.C.

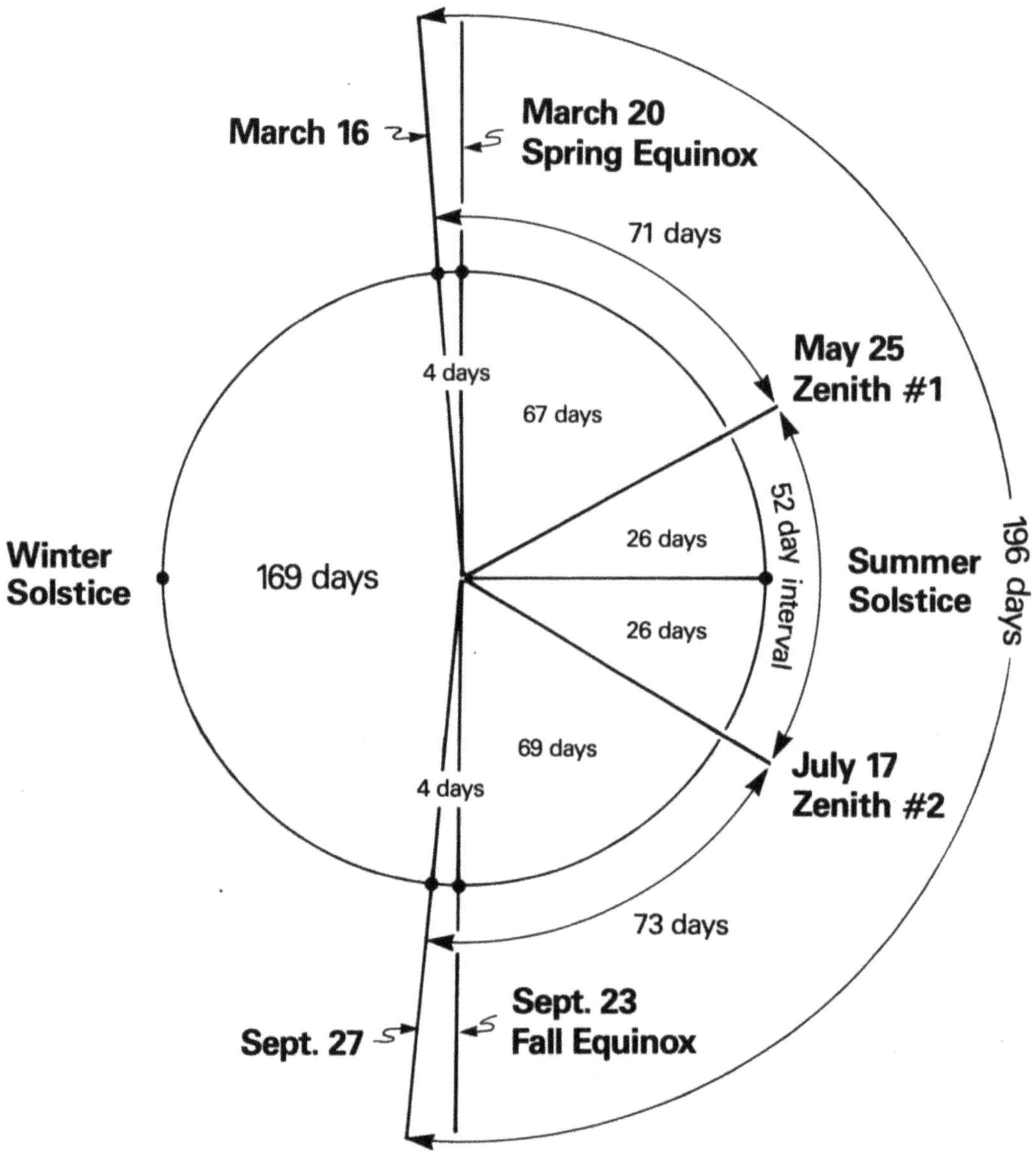

Chart 1. Postulated divisions of the solar year at Dzibilchaltun (21 deg. 5'45'' North latitude).

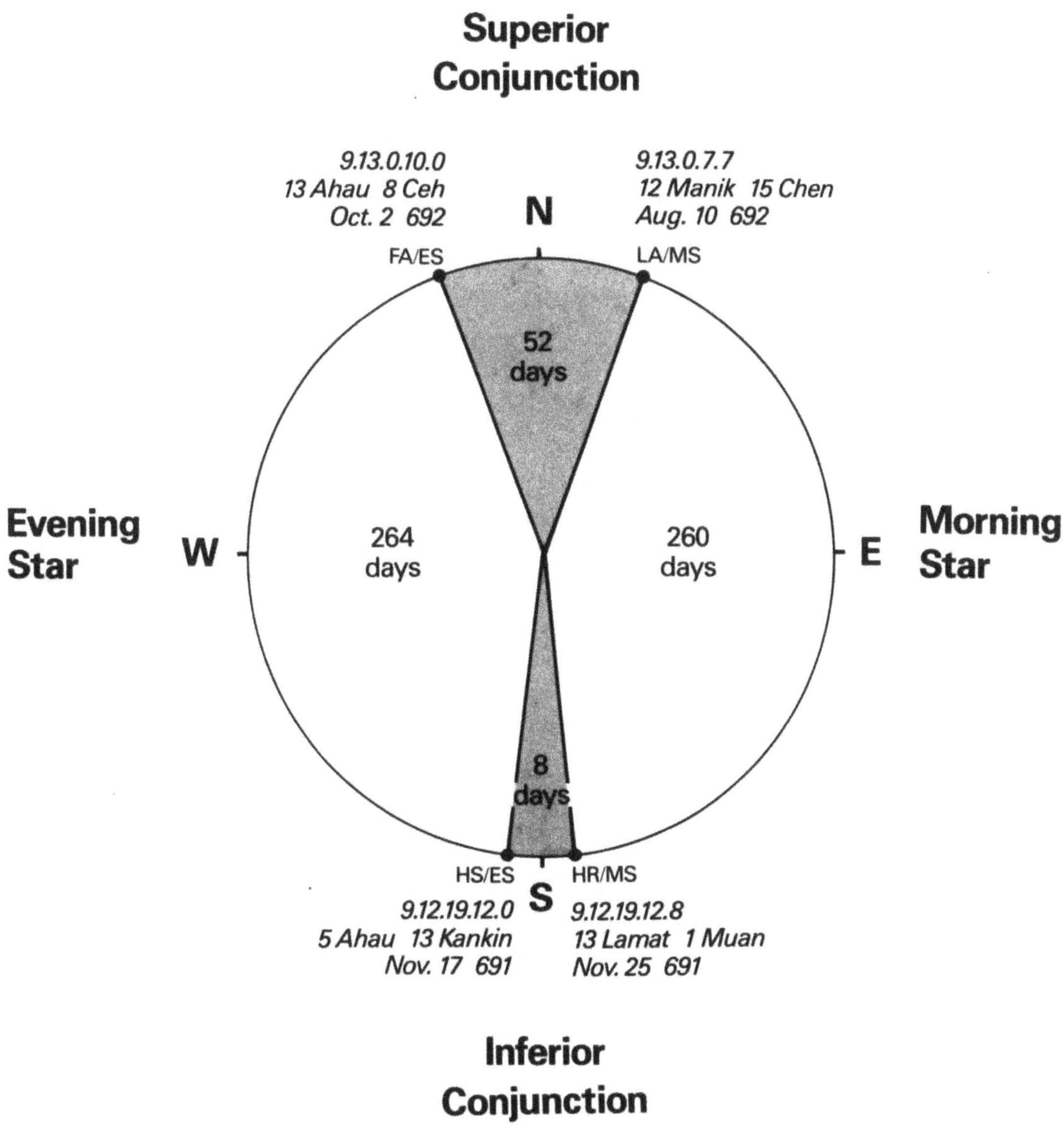

Chart 2. A Canonical Venus cycle at Dzibilchaltun.

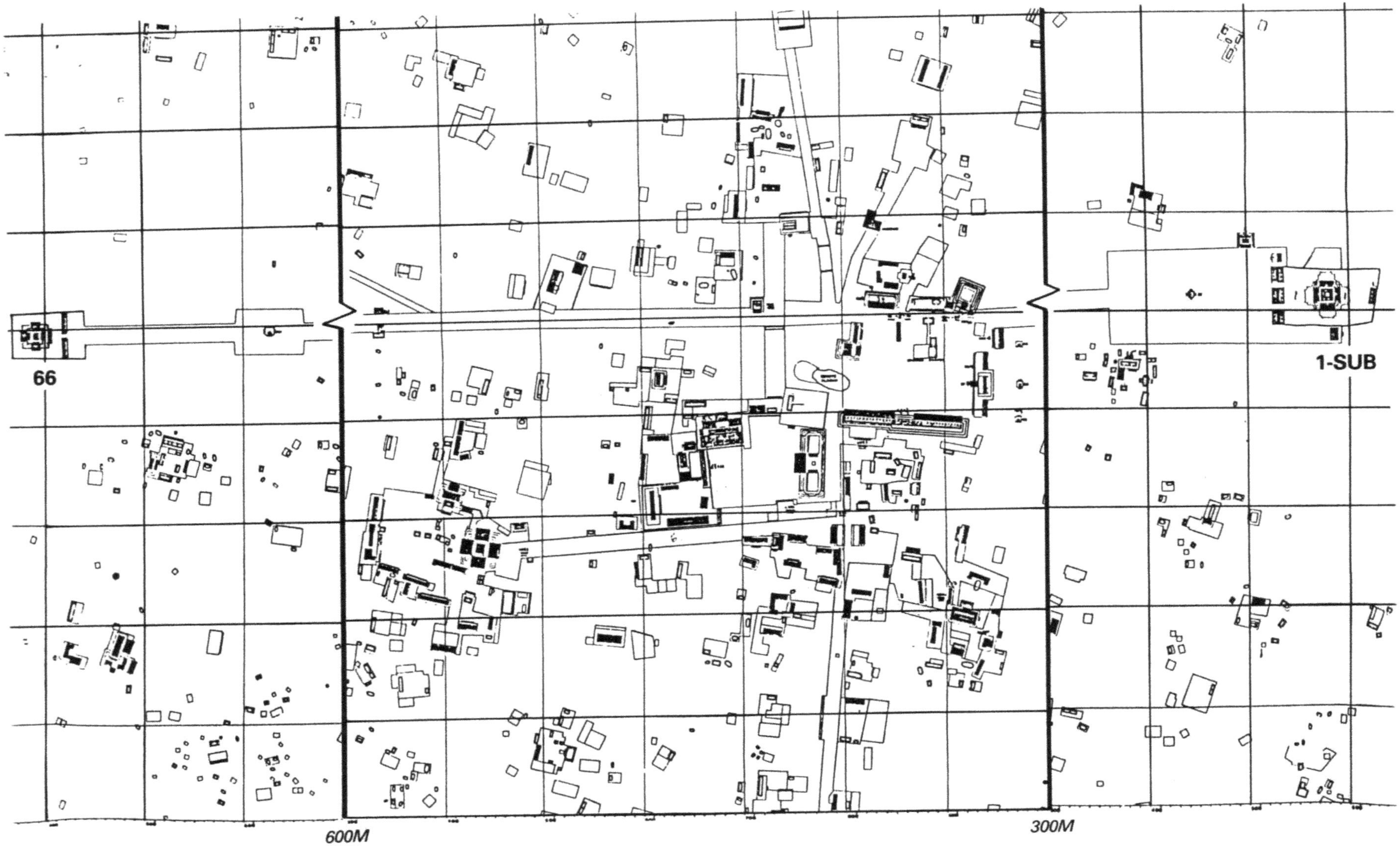

1. Map of Dzibilchaltun, with 900 m deleted from the sacbe (after Coggins 1983, fig. 2).

2. Structure 1-sub viewed from southwest of stela platform for view through the axial doorways of Str. 1-sub. (Courtesy of Horst Hartung).

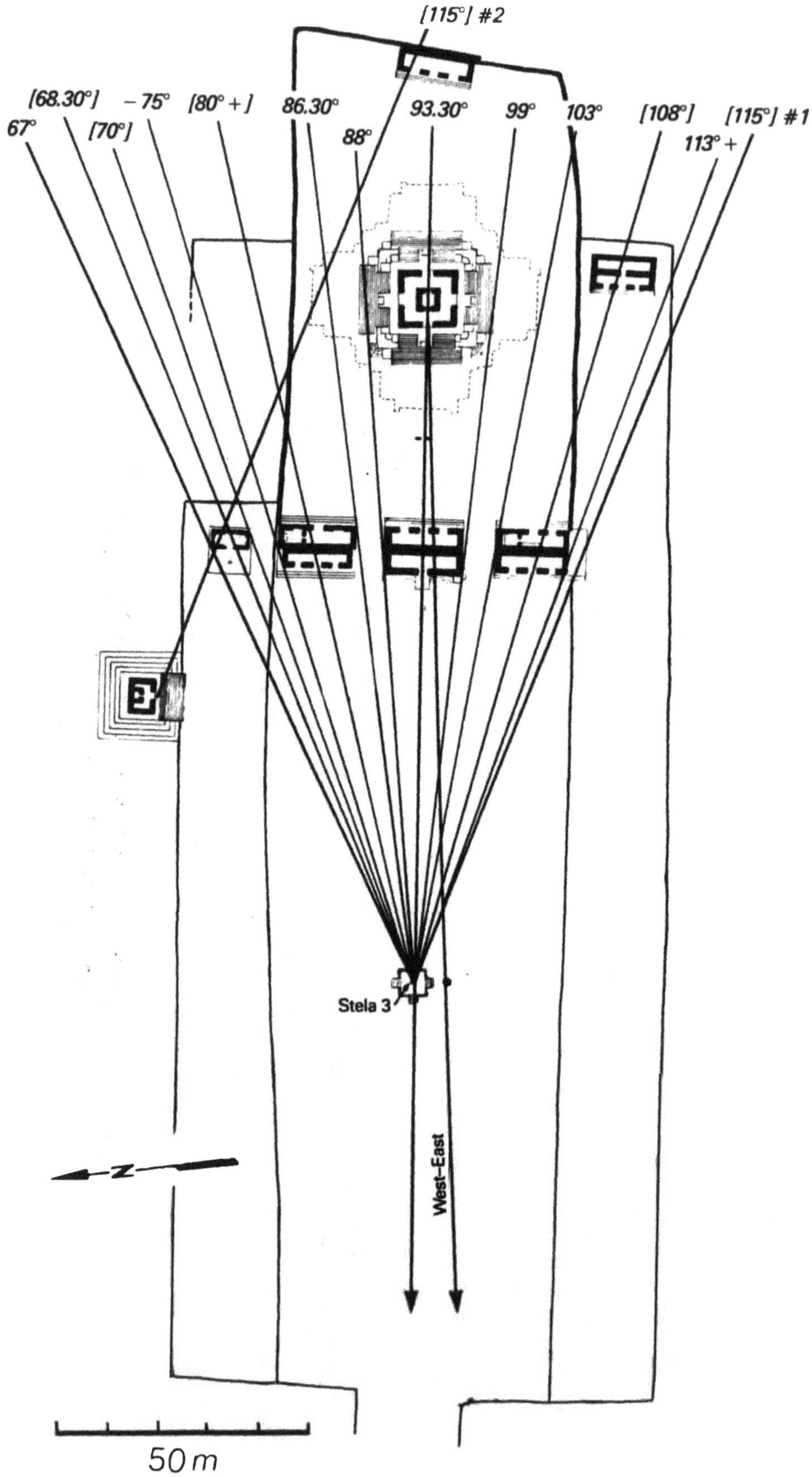

3. Str. 1-sub group with bearings measured from Stela 3, on Str. 1. Azimuths in parentheses are postulated.

4. Reconstruction of the Str.1-sub group. (Drawing by George E. Stuart and Lisa Biganzoli. Courtesy of Middle American Research Institute)

5. Mesoamerican four-part signs that denote the completion of cycles. a. Kan cross b. Kin sign (one day) c. Lamat (Venus) d. Teotihuacan flower emblem (xochitl?) e. Katun completion sign f. Codex Fejervary-Mayer, page 1 (calendar cycle). (From Coggins 1983, fig. 33)

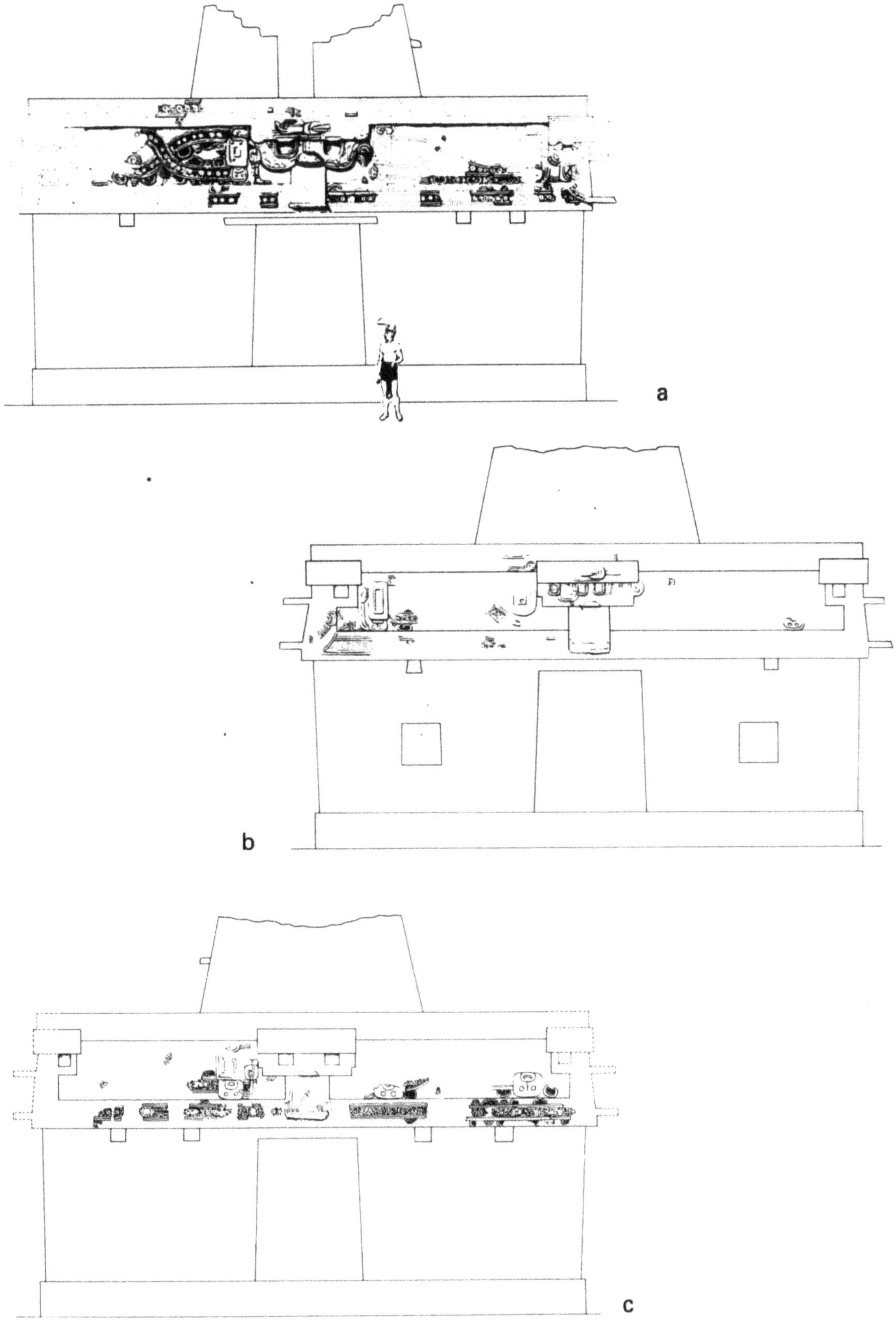

6. Structure 1-sub. Remaining stucco decoration on the
a. South facade b. East facade c. North facade.
(From Coggins 1983, figs. 7-10)

7. Temple of Quetzalcoatl, Teotihuacan. Facade decoration. Drawing by Barbara Page. (From Coggins 1983, fig. 36)

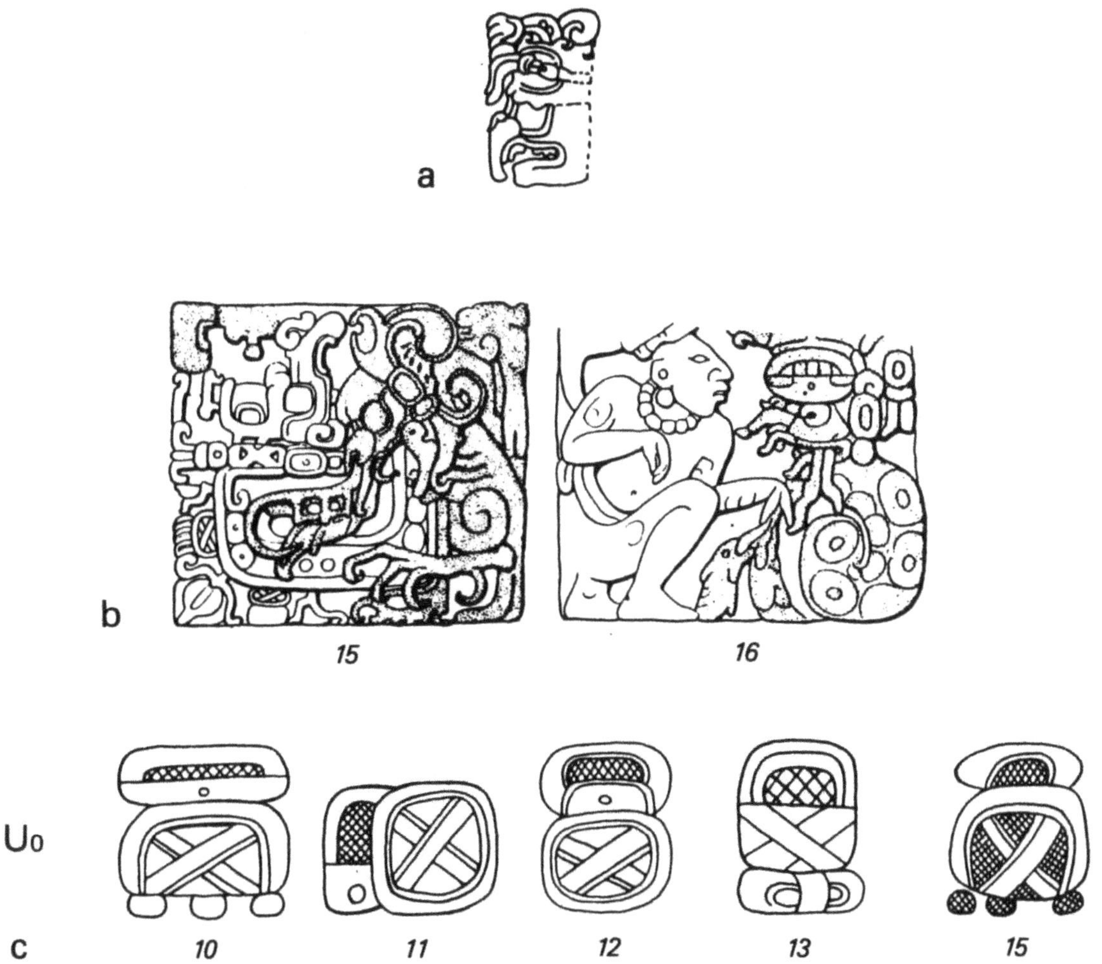

8 a. Head variant of the number thirteen (From Thompson 1960, fig. 25:9). b. Full figure variants of the number thirteen and of the tun. 15. "Thirteen tuns" 16. "Five tuns" (From Thompson 1960, fig. 28: 15,16). c. Glyphs for the month Uo (From Thompson 1960, fig. 16:10-15)

9. Incised bone from Tikal Burial 116 (MT 50). (Courtesy of the University Museum, University of Pennsylvania).

10. Copan Stela N, south face. (From Maudslay 1889-1901, I, pl. 82)

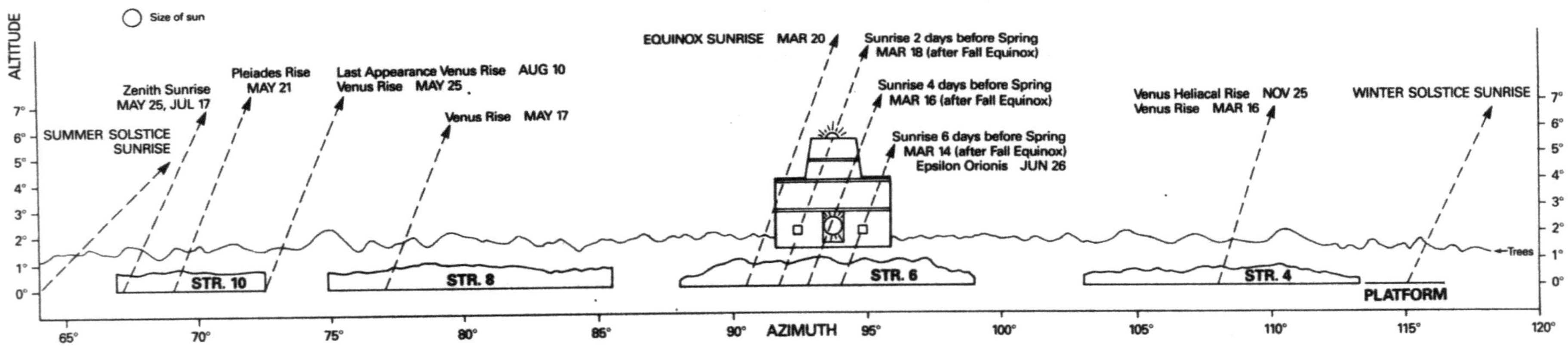

11. Elevation of Strs. 1-sub, 4-10 with solar rise points by Aveni and Hartung; Venus, Pleiades, Epsilon Orionis added. (Redrawn by Nancy Lambert-Brown).

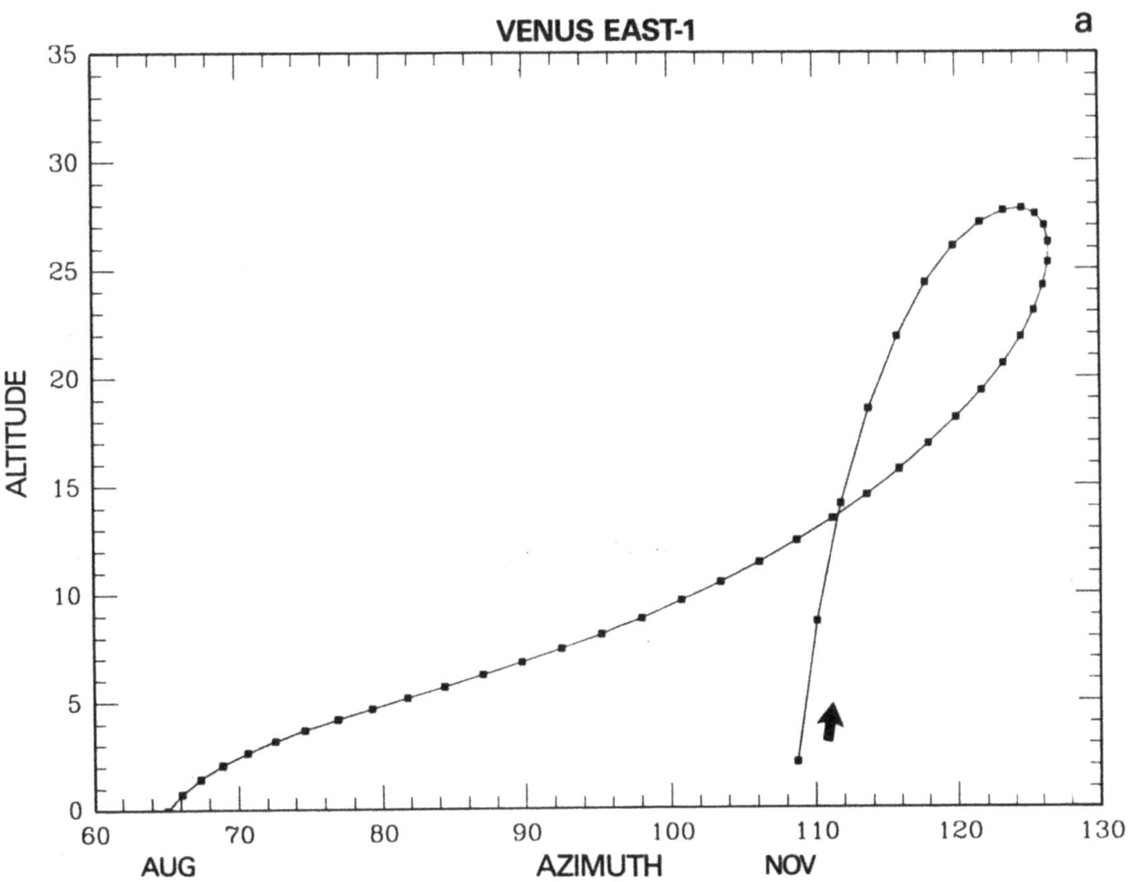

12. The five configurations of Venus as Morning Star. Charts show the greatest height of Venus at five day intervals. The sun is 15 degrees below the horizon on these charts although Venus is visible when as close as 7 degrees to the sun. The amended chart in fig. 13 illustrates the full Venus period (Courtesy of Owen Gingerich, Harvard-Smithsonian Center for Astrophysics). a. Dzibilchaltun cycle 1. November 25-August 10, A.D. 691-2. b. Cycle 2. July-February, 693. c. Cycle 3. February-October, 695. d. Cycle 4. September-April, 696-7. e. Cycle 5. April-December, 698.

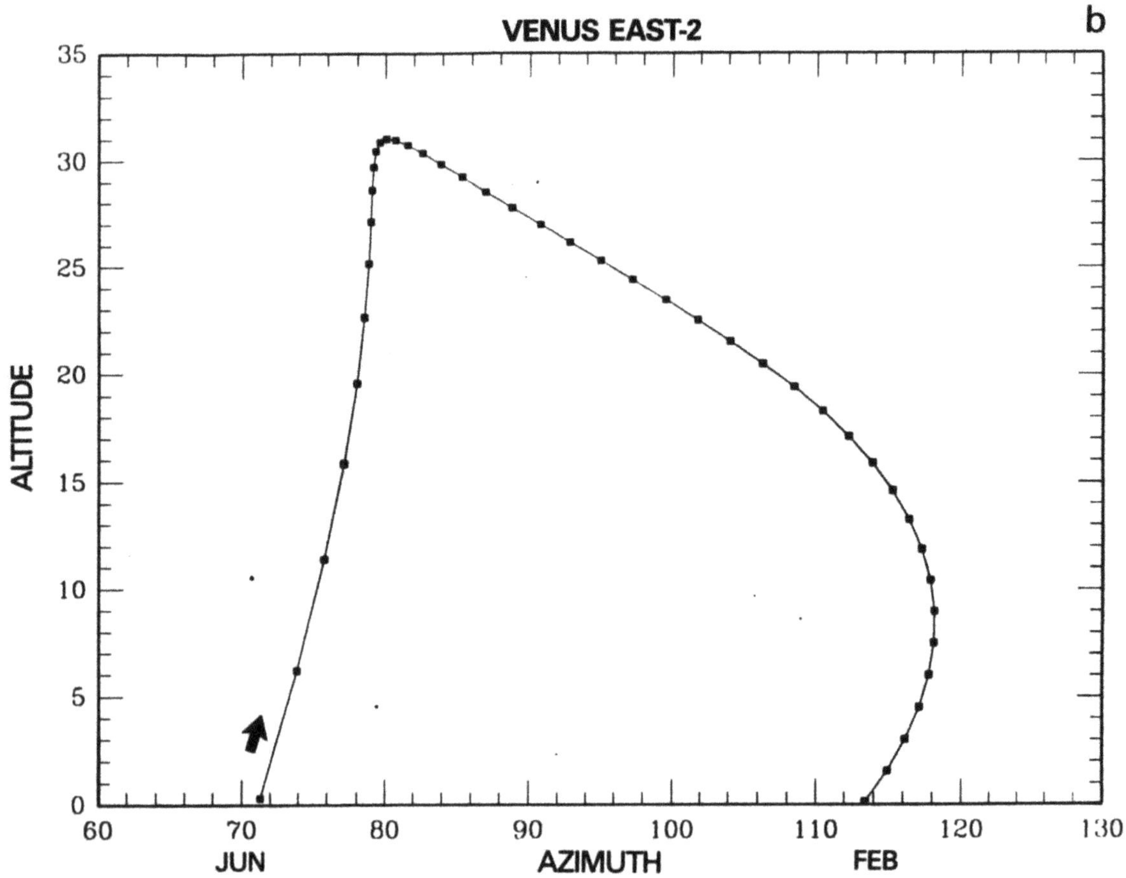
VENUS EAST-2
b
ALTITUDE
AZIMUTH
JUN
FEB

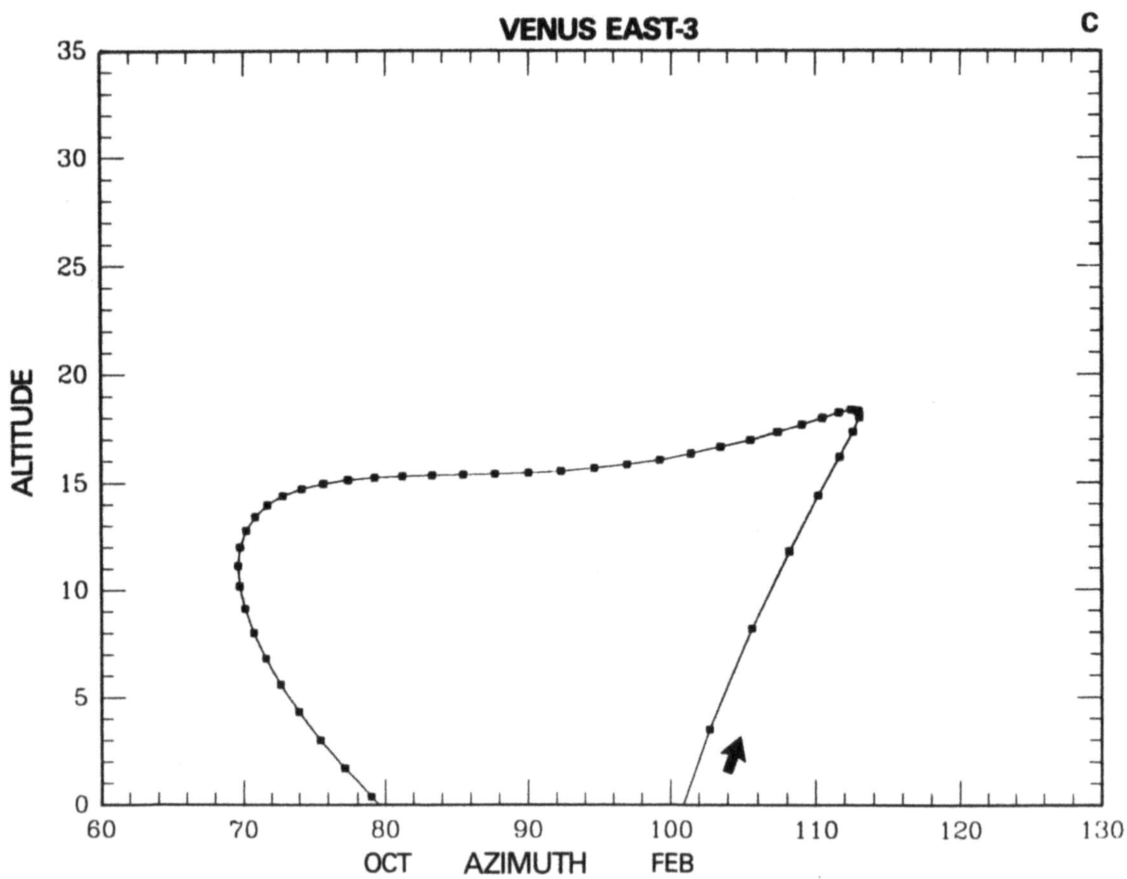
VENUS EAST-3
c
ALTITUDE
AZIMUTH
OCT
FEB

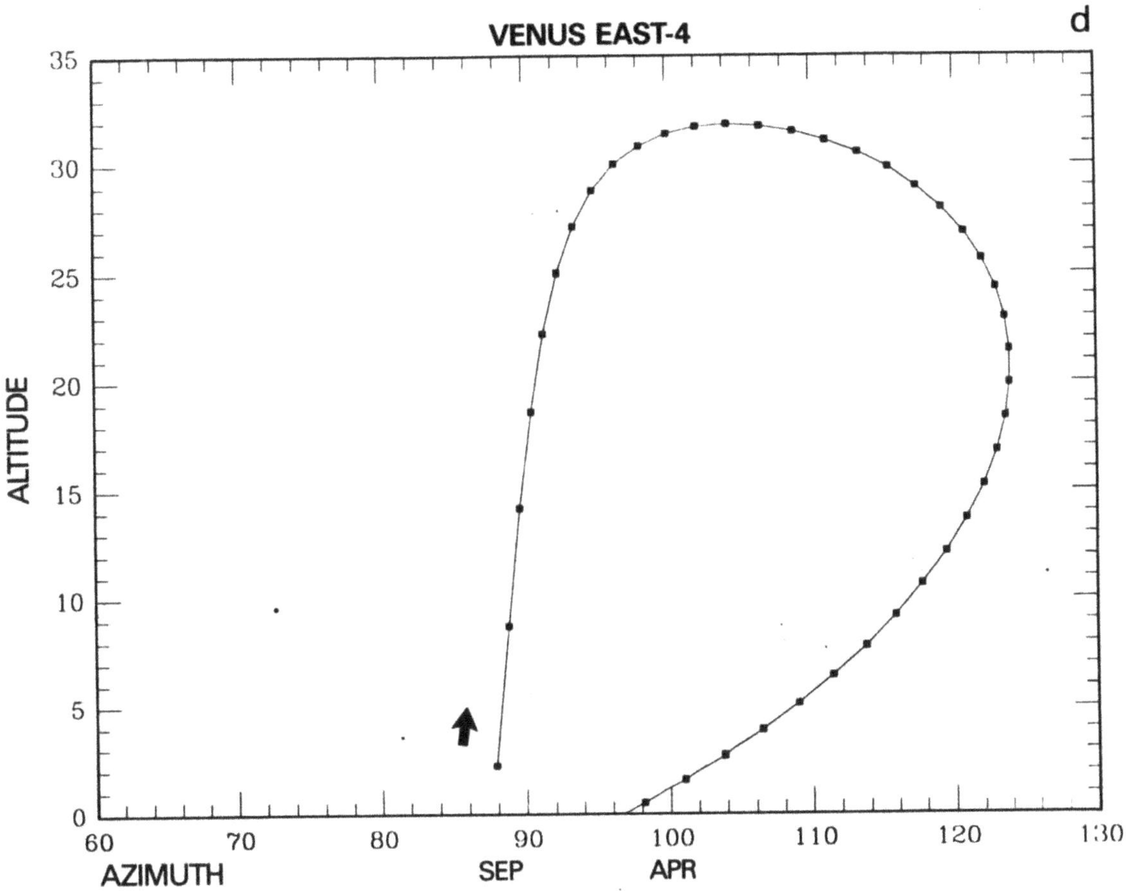
d
VENUS EAST-4
ALTITUDE
AZIMUTH
SEP
APR
0
5
10
15
20
25
30
35
60
70
80
90
100
110
120
130

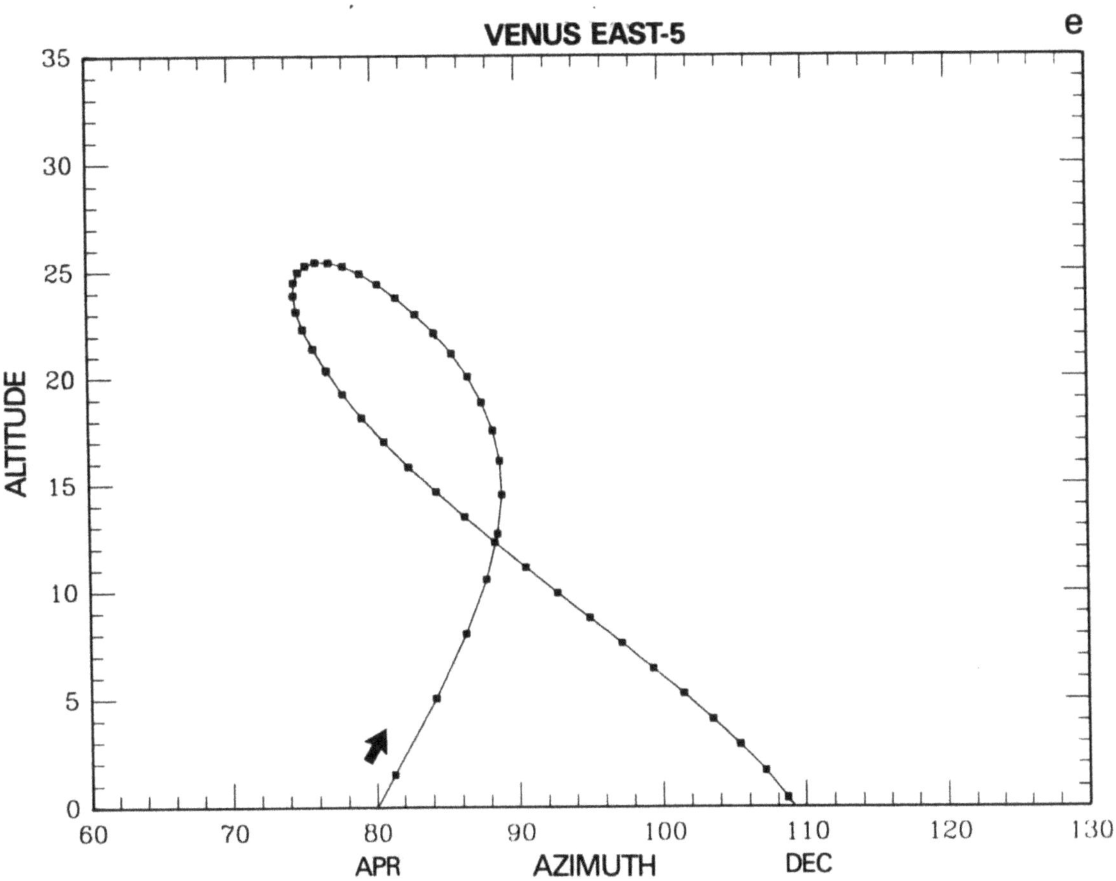
e
VENUS EAST-5
ALTITUDE
AZIMUTH
APR
DEC
0
5
10
15
20
25
30
35
60
70
80
90
100
110
120
130

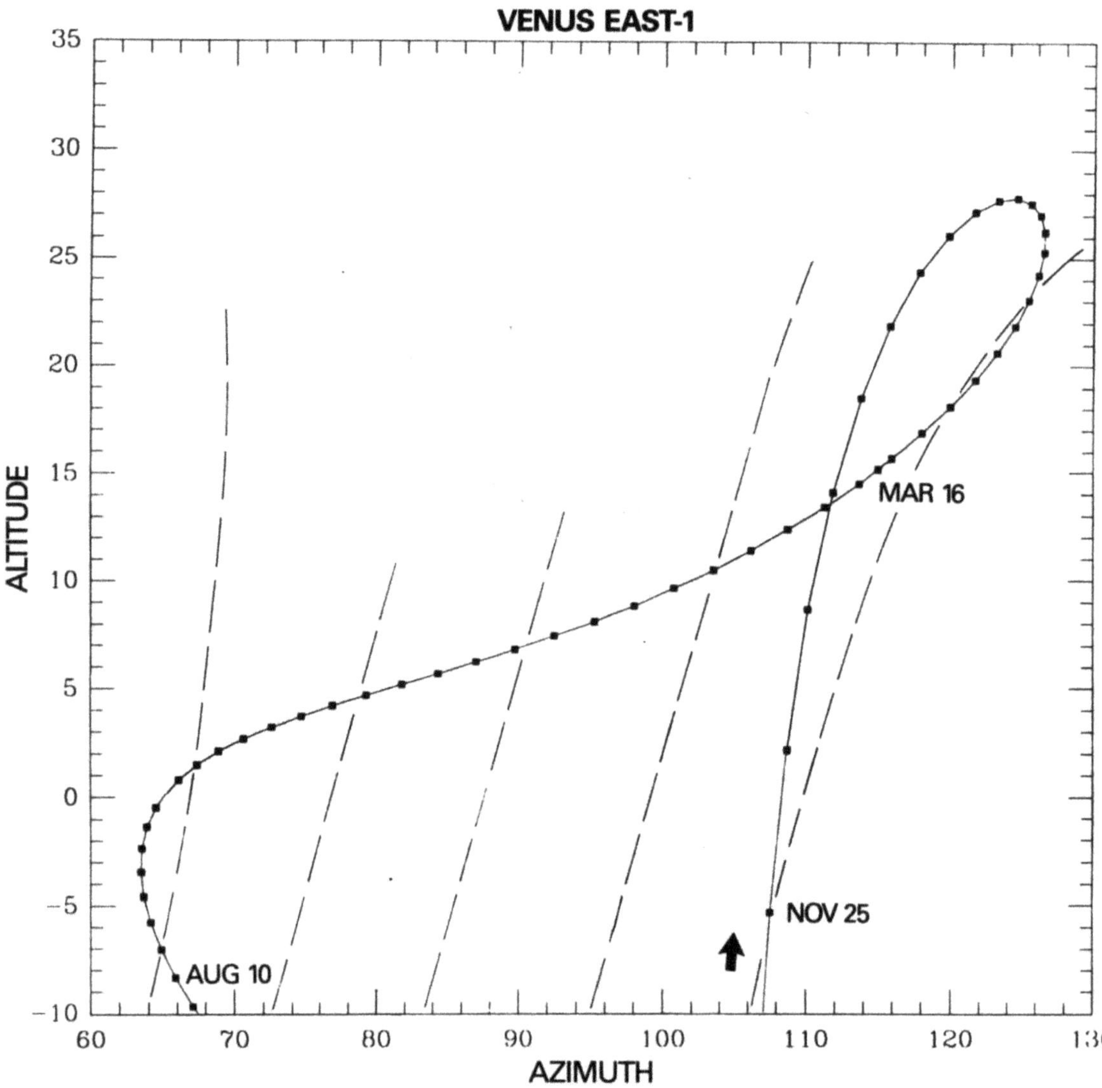

13. Dzibilchaltun Cycle 1. November 25-August 10, 691-2, showing the full cycle. (Dotted lines indicate angle of rise). (Courtesy of Owen Gingerich, Harvard-Smithsonian Center for Astrophysics)

Astronomical Images and Orientations in the Architecture of Chichen Itza

Susan Milbrath

Chichen Itza is one of the most important Mesoamerican sites for study of astronomical orientations in architecture. The Caracol, with its unusual tower windows, has long been recognized as a Yucatec Maya observatory (Aveni, Gibbs and Hartung 1975). Many other buildings at the site have very interesting astronomical orientations that have not been previously recognized. The most important orientations seem to be towards the zenith sunset, the setting of the Pleiades, and the first appearance of Venus as the evening star at the beginning of the rainy season.[1]

Although most previous work on Mesoamerican architectural orientations has concentrated on measurements and corresponding celestial events, it seems that the alignments may also hold keys to understanding the symbolism of individual buildings. At Chichen Itza the astronomical orientations of various buildings do seem to provide important information for the interpretation of murals and relief sculptures.

Site Chronology

The chronology of Chichen Itza is still much debated because there are two disparate styles of architecture at the site: the Puuc or Chichen-Maya style and the Toltec-Maya style. The Puuc style is named for the hilly region of northern Yucatan, where mosaic architectural decoration predominated during the late Classic period (A.D. 600-900). Chichen-Maya style buildings bear calender inscriptions that cover a period of less than 40 years, between 10.2.0.0.0 and 10.4.0.0.0 (Kelley 1982:Table 1). Kowalski (1987:50) notes that the majority of these dates fall in a 12 year period, from A.D. 869-881. According to Tozzer (1957), the Chichen-Maya or Yucatan-Maya period dates around A.D. 700-950 and the Toltec-Maya period dates between A.D. 987-1263, reflecting an Early Postclassic instrusion from the Central Mexican site of Tula. Kubler (1961) suggests that elements of the Toltec style are actually more developed at Chichen Itza than at Tula, indicating Chichen Itza was probably the originator of certain architectural features at Tula.

A number of scholars have argued that the Great Ball Court complex, with its "Toltec" style features, actually correlates with Late Classic Mesoamerican art styles (Parsons 1969, Cohodas 1978, Pasztory 1972). The chronological placement of the Puuc or Chichen-Maya style as earlier than the Chichen-Toltec style is also being questioned, and now some scholars suggest that both styles existed contemporaneously (Cohodas

1978:Chart III; Wren 1986a, 1986b, Lincoln 1986).

Kowalski (1987:238) suggests a foreign group of Toltecs or Mexicanized Chontal Maya came to Chichen Itza, occupying the Itza city in order to establish authority over trading routes that were in disarray due to political rivalries in northern Yucatan. A relalatively small group of elite outsiders could have inspired a new style of architecture blending foreign elements.

The lack of extensive scientific excavations has hampered accurate dating at Chichen Itza. At present there are only a few clear chronological markers, including radiocarbon dates of A.D. 600 ± 70 and A.D. 780± 70 for the Puuc style Iglesia, and dates of A.D. 790 ± 70 and 810 ± 100 from the Chichen-Toltec style Castillo (Andrews and Andrews 1980:Table 4). The architectural inscriptions range between 10.1.17.15.3 and 10.3.5.5.9, A.D. 867-894 in the Thompson correlation, and the latest date in architectural context is 10.8.10.11.0 (A.D. 998), found at the High Priest's Grave (Krochock 1988:152-153, Fig. 19), a structure that may late in the architectural chronology of Chichen Itza (Wren 1986a; Parsons 1969:Table 7).

Wren's (1986a, 1986b) study of Chichen Itza inscriptions and relief sculpture, and her analysis of a circular stone relief suggest the Great Ball Court Complex dates between A.D. 869-889. Wren notes the circular stone, which may have been originally located in the Great Ball Court, has a Maya style Calendar Round date encircling relief figures that resemble murals and reliefs of the Great Ball Court Complex.

A ninth century date for the Great Ball court seems likely, based on the radiocarbon dates for the Castillo and the span of dates from inscriptions. Fortunately, establishing the exact chronology of the buildings is not essential to our analysis. Our discussion of astronomical alignments is based primarily on repeating celestial cycles, such as the solar zenith sunset and the setting of the Pleiades.

Solar Zenith in Yucatan

In Mesoamerica, as in other areas of the tropics, the sun reaches an overhead or true zenith position at noon twice a year (Aveni 1980:66, Table 3). Unlike temperate zones, where there are four distinct seasons correlating with the equinoxes and solstices, in Mesoamerica there are two seasons: the dry season and rainy season. About 80 percent of the annual rain falls between May and October (Vivo Escoto 1964:201). Throughout much of Mesoamerica the first solar zenith marks the transition from the dry season to the rainy season, when planting takes place today, as it did in ancient times (Milbrath 1981: 270, 281).

The solar zenith date varies with latitude. In the area of Chichen Itza the first solar zenith falls on May 25th, and the second solar zenith on July 20th. In

Yucatan the planting begins in May, usually after the rains begin (Trujillo 1946:203). As a marker for both seasonal change and agricultural activity, the first solar zenith may be the single most important solar event in the Yucatec Maya calendar. In ancient times the second solar zenith also had great importance as the beginning date of the ceremonial calendar - the haab (Aveni 1980:42-43; Tozzer 1941:151-153).

The zenith position has special significance because the sun seems to be closest to the observer, in the "center" as it were. The Maya, with their structured calendar and long distance trade patterns, were aware that the sun reached zenith at varying times of year in different latitudes as they traveled north and south. It is probable that each Maya site was distinguished by its own special ceremony, when the sun reached zenith at an individual local. The overhead position of the sun could be easily determined by watching the changing length of the noontime shadows, with the zenith being the day the sun casts no shadow at noon. Marking the sun's horizon position from a fixed observation point on that date would allow the priests to regulate the agricultural calendar.

Zenith Orientations at Chichen Itza

With a completely flat horizon, the zenith sunset at the latitude of Chichen Itza (at 20° 40′ latitude north) is located at 22° north of west, correlating with an azimuth of 292° (Aveni 1980:Table 9, 1972; Table 1a). An azimuth is our modern measurement incorporating the concept of a 360° circle, starting with the zero point at true astronomical north. Ancient Mesoamerican astronomers did not use this kind of reckoning, and they determined alignments using horizon features and simple sighting devices like crossed sticks (Aveni 1980:Fig. 26). Once the alignments had been established they were often codified in architectural orientations.

Owing to the flat horizon in the area of Chichen Itza, fixing specific horizon points for observation was especially difficult. Given the featureless horizon and the lack of precision instruments for astronomical alignments, it may be that architectural orientations were sometimes off by a degree or even two. Nevertheless, precise orientation is evident in the Caracol, a building designed specifically as an observatory (Aveni, Gibbs and Hartung 1975).

The solar zenith may have been especially important in the architectural orientations at Chichen Itza. The Caracol observatory has a stairway niche with one wall precisely oriented to the zenith sunset at an azimuth of 292° or 22° north of west and the other wall aligned to the maximum northern setting position of Venus (Aveni 1980:Fig. 90). The Castillo, a building with special calendrical significance, has a western orientation of 291°, only one degree from the solar zenith sunset (Fig. 1; Aveni 1980:Table 9, Appendix A).

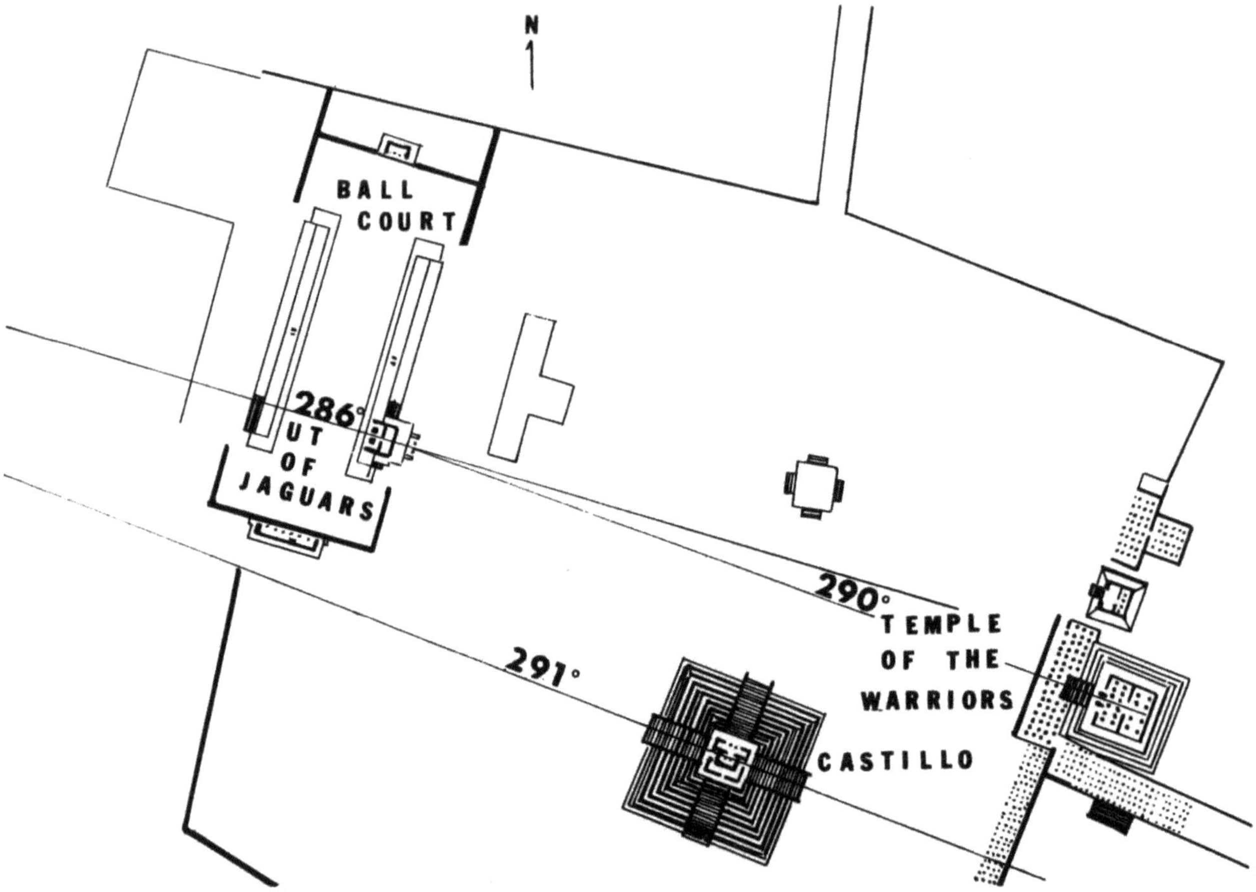

Fig. 1. Map of the astronomical alignments of the North Terrace Group of Chichen Itza, Yucatan, Mexico.

Fig. 2. The Upper Temple fo the Jaguars overlooking the Great Ball Court. The Castillo is in the background.

The Castillo (Fig. 2, background) has exactly 365 steps in the four radiating staircases and temple platform. The pyramid has been the source of great interest because the unusual play of sunlight and shadow at sunset on the equinox creates a pattern along the serpent balastrade that mimics the markings of a diamond-back rattler (Rivard 1969; Krupp 1983:298-299). Coggins (1986) believes the pyramid was constructed to commemorate the completion of the 10th baktun on March 13, 830, a date close to the spring equinox.

Astronomical alignments of the Castillo reflect the fundamental division of the Yucatec Maya year into two seasons or halfs. The west face with its orientation towards the setting sun at the solar zenith (Fig. 1), reflects the dividing point between the dry season and the rainy season. The solar zenith in late May is when the rains begin, which signals the first planting today, as it did in ancient times in Yucatan (Trujillo 1946:203; Milbrath 1981:281).

The east facade of the Castillo, with an orientation of 111° (180° rotation from the east face; Fig. 1), faces the rising sun six months after the solar zenith. This is the time of the solar nadir (November 22) at the latitude of Chichen Itza. The eastern orientation, facing sunrise on the day of the solar nadir, marks the position of the sun at the transition from the rainy season to the dry season in late November.

November is the month the dry season and the maize harvest begins in Yucatan (Vivo Escoto 1964:201; Perez Toro 1942; Trujillo 1946:203). The Mesoamerican maize harvest is a slow process because many of the doubled-over maize ears are allowed to dry on the stalks, and there are two maize crops to harvest, the early maize planted in May and the late maize planted in June. In January the late maize is harvested and work is begun on clearing the fields in preparation for burning (Trujillo 1946:203). January is also the month of the second solar nadir (January 21; Milbrath 1981:279-280). Hence the long maize harvest is bracketed by the two solar nadirs.

The Solar Nadir

Like the solar zenith, the solar nadir is a phenomenon confined to the tropics. The solar nadir is the time when the ecliptic, the path of the sun and planets, stands upright at midnight like a giant wheel with the zodiacal constellations arching overhead. At this time the sun is 180° away, directly below the earth at the latitude of the observer in the tropics.

The solar nadir can be calculated by watching the stars moving opposite to the sun. During the Terminal Classic and Postclassic periods, in the latitudes of the Valley of Mexico and northern Yucatan (which are located approximately at parallel latitudes), the midnight zenith of the Pleiades correponded approximately to the time of the November solar nadir, when the sun was in opposition to the Pleiades located 180° away. At this time the sun crossed the Milky Way in the region of Scorpius, at the intersection of the Milky Way and the ecliptic. This cosmic crossroads in the sky mirrors the other intersection point of the Milky Way and the ecliptic in the region vicinity of the Pleiades (Milbrath 1980; 1981: 281).

According to Friar Sahagun, every fifty-two years the the Aztecs celebrated the New Fire Ceremony as a sign of world renewal. Sahagun (1950-1982:IV, 143) informs us that the ceremony was timed by the midnight zenith of the Pleiades in November. More recently, scholars have recognized that the ceremony also celebrated the solar nadir (November 20) at the latitude of the Aztec captical of Tenochtitlan (Milbrath 1980; Krupp 1981). We know the Yucatec Maya told time by observations of the Pleiades, and they must also have been aware of the opposition of the sun and Pleiades at the time of the November solar nadir (Milbrath 1980, 1981).

Although we have no direct proof that the solar nadir was recognized among the Yucatec Maya, the architectural orientation of the east facade of the Castillo provides evidence for observation of this solar event. Solar ceremonies may have been performed in the eastern chamber of the Castillo at the time of the solar nadir, when the sunlight at dawn penetrated the dark recesses of the temple.

<u>Lower Temple of the Jaguars</u>

The Lower Temple of the Jaguars has an orientation of 106°, close to the direction of the rising sun at the time of the solar nadir (Fig. 1). The actual orientation is towards sunrise several weeks before the solar nadir in late November and several weeks after the solar nadir in late January. Elaborately carved scenes on the interior of the temple receive the most amount of direct sunlight at sunrise in early November and early February.

Painted reliefs in the interior temple depict five rows of warriors and deity figures, and a prominent sun disk in the uppermost register (Cohodas 1978:220-225, Figs. 16-18; Marquina 1964:Fotos 435, 436). The building may be equated with the Aztec Tlacochcalco, which functioned as an armory and military headquarters (Wren 1986b). Warrior cults and solar images depicted in the upper register indicate a military cult linked to sun worship. The orientation suggests the reliefs represent militaristic ceremonies performed during the dry season, when the rising sun would penetrate the temple and bring the warriors to life.

Both annual solar nadirs coincide with the dry season, bracketing a period of less intense agricultural activity in Yucatan when warrior cults may have been emphasized. Early Yucatec Maya ethnographic sources note that warfare was usually confined to the dry season (Reed 1964:99). Similarly, ethnohistorical documents inform us that the Aztecs timed all their major military campaigns to coincide with the dry season because roads were most easily traversed at that time (Hassig 1988).

Fig. 3. Feathered serpent columns in the Upper Temple of the Jaguars.

Upper Temple of the Jaguars

Looming over the largest ball court in Mesoamerica, the Upper Temple of the Jaguars has a clear view of the flat horizon (Fig. 2). As in the Temple of the Warriors, entry to the interior space is through twin feathered serpent columns (Figs. 3, 4). Because the inner temple has been preserved intact, the architectural alignment today remains as it was in the past. The temple has an orientation of 286° (16° north of west), slightly skewed from the actual zenith sunset.

Fig. 4. The solar zenith sunset illuminates the interior of the Upper Temple of the Jaguars.

The positioning of the Upper Temple of the Jaguars creates an interesting play of light and shadow around the time of the solar zenith sunset. At sunset a rectangle of sunlight penetrates the inner chamber like a spotlight passing in between the feathered serpent columns, lighting the jambs and back wall of the inner chamber (Fig. 5). The inner temple was painted with colorful murals that have all but disappeared. Fortunately, they were recorded in the 19th century by Adela Breton (Coggins 1984:Figs. 17-20). The spotlit murals would have added dramatic effects to rituals performed at sunset in the Upper Temple of the Jaguars around the time of the solar zenith.

Prior to modern tree growth atop the west side of the ball court, the last rays of the setting solar zenith sun would have illuminated the carved wooden lintel of the inner temple. On both the outer and inner faces, the lintel depicts a feathered rattlesnake and a sun disk flanking a water-lily god, but only the inner

Fig. 5. Light penetrates the inner temple, lighting jambs and back wall.

face has a Venus glyph positioned over the water-lily (Maudslay 1889-1902, v. 3, Pl. 35; Cohodas 1978:Fig. 39). Both faces show two human figures, one with the feathered serpent and the other seated in the sun disk. These figures have been identified as Captain Serpent and Captain Sun Disk (Miller 1977; Coggins 1984:163). More recently Lincoln (1988) proposes these figures are Kukulcan (Sprouting Serpent) and Kakupacak (Fire

Shield), dual rulers of Chichen Itza. Even though they may be historical figures, the prominence of the sun disk and the Mesoamerican Venus god, the feathered serpent (Aveni 1986), indicates a link with the celestial realm. A similar blend of astronomy, history and politics is evident in Aztec monumental sculpture (Milbrath 1983).

Fig. 6. Mural on rear wall of the inner temple.

When the murals in the Upper Temple of the Jaguars were still intact, two deified rulers in the center panel of the east wall were illuminated around the time of the solar zenith (Fig. 6). According to current interpretations, the figure on the left is Captain Sun Disk, a Maya ruler associated with the sun disk, and the right hand figure is Captain Serpent, a Mexican or Toltec ruler affiliated with the feathered serpent (Miller 1977; Coggins 1984: 160). Astronomy and history may be fused if the figures are in fact co-rulers who claimed celestial patronage or heavenly ancestors, perhaps even the legendary founders of Chichen Itza.

Kelley (1982:2-5) suggests Kakupacal, a person frequently mentioned in the Maya inscriptions at Chichen Itza, is the Itza captain of the chronicles and the founder of Chichen Itza. He believes the murals of the Upper Temple of the Jaguars show a Toltec group conquering the Itza, creators of the Puuc architecture at Chichen Itza. He identifies Quetzalcoatl and Mixcoatl, the leaders of the Toltec army, but he makes no specific identification of the conquered Itza lord in the mural scenes. Miller (1977) suggests the conquered lord is Captain Sun Disk, whom Lincoln (1986) identifies with Kakupacal. It seems more than a coincidence that this ruler's name could be read as a name for the sun itself. We may have another example of a divine "sun king" among the Maya (Milbrath 1983).

The dynamics of politics and astronomy are developed further in a complex sequence of battle scenes. Coggins (1984) agrees with Miller's (1977) historical interpretation, but adds that Venus and solar positions are recorded in the murals. The scenes of combat in the Upper Temple of the Jaguars probably reflect a link between Venus, astrology and warfare also found in the Classic Maya murals of Bonampak (Lounsbury 1982; Schele and Miller 1986:216-217, Fig. V.6; Coggins 1984:157). The emphasis on astronomical symbolism in the architecture does not invalidate current interpretations relating to historical events (Lincoln 1988; Kelley 1982; Wren 1986a, 1986b), but rather it adds a rich, new dimension. Like the Classic Maya, the builders of Chichen Itza chose to link astronomy and history, most probably to set historical events within a grand scheme related to cosmic events.

The murals in the Upper Temple of the Jaguars may relate to the Venus cycle. Coggins argues that the murals represent both the division of the day into eight parts and an eight-part division of the 584 day Venus cycle, starting with Venus as the morning star before dawn, and following through to the end of the sequence in the middle of the night with Venus in inferior conjunction. The daily motions of the sun are linked conceptually with the 584 day Venus cycle, according to her interpretation. Her analysis combines two unrelated cycles and does not represent a real astronomical situation. Coggin's identification of an eight part mural sequence, however, does provide an intriguing correlation with the Venus cycle. The 365 day year and the 584 day Venus cycle coordinate in such a way that every eight years Venus returns to the same position at the same time of year. We know the Postclassic Yucatec Maya recorded the eight year Venus cycle in the Dresden Codex (Aveni 1980:186). This cycle also may be depicted in the eight mural scenes of the Upper Temple of the Jaguars, with each scene showing Venus in a different position in the sky. The exact patterns that emerge over an eight year period will be discussed below.

The Pleiades and Venus

If observations of Venus were made over the course of eight years, the astronomers may have selected a specific phase of Venus, observing that phase in relation to a specific season, horizon position or stellar coordinate in order to measure Venus positions. The first appearance of Venus as the evening star recurs once every 584 days; but the first appearance of Venus as the evening star during a specific season would occur only once every eight years. Similarly, rising and setting positions of Venus would repeat in a pattern every eight years (Aveni 1986). Interest in the extreme northern setting position of Venus in relation to the solar zenith sunset is evident in the orientation of the

Caracol stairway (Aveni 1980:Fig. 90).

If a stellar coordinate were selected, Venus would be positioned close to a specific star about once a year, but only every eight years would Venus be in a specific phase as it approached the stellar position. The most likely candidate for a stellar coordinate is the Pleiades, which were observed to tell time in Postclassic Yucatan (Tozzer 1942:133).

The Pleiades were called tzab or "the rattlesnake's rattle" by the Yucatec Maya at the time of the conquest (Tozzer 1941:133). The primary language of the residents of Chichen Itza probably was Yucatec Maya (Edmonson personal communication, 1988), which provides a linguistic basis for interpreting the rattlesnake capitals on the serpent columns at Chichen Itza as representations of the Pleiades. This interpretation is reinforced by the fact that the Upper Temple of the Jaguars is orientated towards the position of the Pleiades on the western horizon. The temple's alignment is almost identical to that of Window 1 of the Caracol, which had a midline facing sunset on April 28, with the right side of the window framing the Pleiades setting as they disappeared for a period of a month on that date (Aveni 1980:Fig. 93; Aveni Gibbs and Hartung:Table 1, Fig. 10).[2] Similarly, the wide door of the Upper Temple of the Jaguars framed the Pleiades as they set one month before the solar zenith on May 25th.

Pleiades orientations are evident in other structures with feathered rattlesnake columns, such as the Temple of the Warriors (Fig. 1) and the Temple of the Wall Panels (Retablos; Aveni 1980:Appendix A; Morris et al. 1931; Ruppert 1931). Both of these building have images of the feathered serpent repeated in relief sculptured panels, and they are both aligned towards the setting position of the Pleiades at 290° (with a 0° horizon around A.D. 1000; Aveni 1980:Table 9). This orientation is also close to the position of the zenith sunset (292° with a 0° horizon Aveni 1980: Table 9), but, given the emphasis on the feathered serpent, solar alignment in the architecture may be secondary to the observations of Venus and the Pleiades discussed below.

The Pleiades were probably especially important in the Late Classic and Early Postclassic period because in northern Yucatan they rose at dawn around the time of the solar zenith in late May, after about a month of invisibility. By the Late Postclassic period their setting position also marked the horizon position of the zenith sunset. The Pleiades setting azimuth at 21° latitude north was 289° around A.D. 500. As the end of the Classic period approached, the Pleiades moved closer and closer to the position of the zenith sunset. Around A.D. 1300 they set at 292°, the precise position of the zenith sunset in northern Yucatan (Aveni 1980:Table 9).

The feathered serpent columns at Chichen Itza may relate to an image of Venus as the evening star in conjunction with the Pleiades. The plumed serpent is a widely recognized image of Quetzalcoatl, the god of the planet Venus in Mesoamerican iconography (Piña Chan 1977; Aveni 1986). An image of Venus wearing the Pleiades, the "rattlesnakes's rattle," might refer to a calendric link between the two astronomical entities.

Looking through astronomical tables compiled by Tuckerman (1964) and comparing the longitudes of Venus with the position of the Pleiades in the Classic and Postclassic periods, there is a very interesting and previously unrecognized correlation between the conjunction of Venus with the Pleiades and the onset of the rainy season in Mesoamerica. Venus moved into conjunction with the Pleiades in a repeating pattern that correlated with the beginning of the rainy season. In its circuit through the background of stars, Venus moved to the position of the Pleiades once a year, and invariably this occurred between mid March and mid June - in other words - around the beginning of the rainy season (Table 1).

Since the Pleiades disappeared in conjunction with the sun near the end of April during the Late Classic and Early Postclassic periods (Aveni 1980:Table 10), the evening star in conjunction with the Pleiades would be visible only in late March through late April, just prior to the onset of the rainy season. And the morning star in conjunction with the Pleiades would be visible only in late May through early June, at the beginning of the rainy season. In either morning or evening star phase, Venus near the Pleiades symbolized the rainy season.

Tracing the dates of the conjunction of Venus with the Pleiades in relation to dates of the first and last appearance of the evening star between A.D. 869-889, the suggested dates for construction of the Great Ball Court Complex (Wren 1986a, 1986b), indicates there is an eight year pattern in relation to the specific dates on which the evening star appears or disappears (Table 2). Table 2 also details the Venus phases in relation to conjunction with the Pleiades, showing that, during the 20 year period studied, Venus was in inferior conjunction near the Pleiades only on two occasions, in the years A.D. 877 and 885. This indicates the eight year pattern was not precise, especially in relation to the short inferior conjunction phase.[3]

A most interesting pattern is evident at eight year intervals in A.D. 869, 877 and 885, when Venus remained very close to the Pleiades around the time of their conjunction with the sun (April 20-May 23, A.D. 1000; Aveni 1980:Table 10) at the beginning of the rainy season. In A.D. 877 this was especially dramatic because, due to retrograde motion, Venus was in the region of the Pleiades for almost three months, and during this period it made the transition through three phases, from the evening star, to its disappearance in inferior conjunction, and reappearance as the morning star.

The years A.D. 870, 878, and 886 also provide an interesting pattern. During these years Venus first appeared as the evening star crowned by the Pleiades around the spring equinox (March 21). As the evening star rose higher the Pleiades descended closer to the horizon each night, crossing the path of the evening star on April 12 and disappearing in late April, around a month before the solar zenith and the onset of the rainy season. Venus continued its ascent during the rainy season, until it reversed course and descended to make its last appearance as the evening star around the time of the solar nadir at the beginning of the dry season. The Castillo, with its precise orientation to the solar nadir, and the Lower Temple of the Jaguars, with its approximate orientation to the dry season sunrise, may have been used to observe Venus every eight years in its first appearance as the evening star at the beginning of the dry season.

These recurring celestial events may have been a focus for the orientation of the Upper Temple of the Jaguars, which is aligned facing 16° north of west. This orientation may not be towards the sun setting one month before the rainy season, but rather to the evening star and the Pleiades just above the horizon in April. The iconography of the Upper Temple of the Jaguars, emphasizing the feathered rattlesnake, may represent Venus in conjunction with the Pleiades as a symbol of the onset of the rainy season.

Pleiades-Venus Orientations in Mesoamerica

Other buildings at Chichen Itza reflect a pattern of imagery and orientation that can be linked to Venus and the Pleiades. The Temple of the Warriors and the Temple of the Wall Panels have feathered serpent columns like the Upper Temple of the Jaguars, and their alignments of 290° may reflect an orientation towards the Pleiades setting and Venus appearing or disappearing as the evening star at the beginning of the rainy season. The claw-footed Venus monster in the panels of the Temple of the Warriors may be images of Venus in inferior conjunction (Cohodas 1978:Fig. 64; Klein 1976:10).

Šprajc (1988) analysizes the rain-maize complex in relation to Venus imagery and building orientations. He found a connection between rainy season imagery and the evening star. He notes that over many millenia Venus extremes visible on the western horizon (maximum northern and southern horizon positions) coincided with the beginning and end of the Mesoamerican rainy season. Only a few architectural alignments reflect orientations to Venus extremes, but all the examples are related to the evening star horizon extremes, rather than the morning star positions (Šprajc 1988). Our discussion above suggests Pleiades alignments also should be considered in relation to the evening star-rain-maize complex.

Copan Temple 22, with its unusual window and Bicephalic Serpent bearing both Venus and solar imagery, is an obvious candidate for study of orientations. Anaylsis of the window alignment suggest that the Maya observed the evening star as it made its first appearance in the year prior to a great extreme, when Venus reached a declination that exceeded the limits of the solstice sunset. Similarly, in the year following a great extreme Venus appeared in the window just prior to disappearing in inferior conjunction. The pattern of the first and last appearances bracketing a Venus extreme repeated at eight year intervals, and invariably occurred between April 25 and May 3rd, in other words, at the beginning of the rainy season. The orientation of the window also mirrors the baseline created by the alignment of Stela 10 and 12, marking the position of the sunset on April 12 and September 1 (279° azimuth, Aveni 1980:240; Aveni 986; Closs et al. 1984). The April date falls about a month before the onset of the rainy season.

The dates for the disappearance of the Pleiades can be estimated for the mid 8th century, the probable period for construction of Temple 22, based on a 15 latitude location.. The date for the last appearance of the Pleiades was April 9 circa A.D. 500 circa (Aveni 1971) and several days later around A.D. 750, which would seem to suggest a correlation with the April baseline. As Šprajc (personal communication) points out, however, the Pleiades would not have been visible through the window around A.D. 750 because they set behind the local horizon at an azimuth of 289°, whereas the greatest viewing azimuth allowed by a diagonal sight line through the window is about 285°. This indicates the Pleiades were probably not involved in the Copan window orientation. On the other hand, a number of Mesoamerican temples do seem to reflect orientations towards the Pleiades setting.

The Late Classic Temple of the Plumed Serpent at Xochicalco faces an azimuth of 287°, or 17° north of west (Aveni and Gibbs 1976:513). Around A.D. 500 at the latitude of Xochicalco (18° latitude north), the Pleiades set at 288° with a 3° horizon. Given the pattern linking the Pleiades and the evening star to the onset of the rainy season, it seems likely the orientation and iconography of the pyramid relate to Venus and the Pleiades setting at the beginning of the rainy season.

Teotihuacan, the most important Central Mexican site of the Classic period, has a pyramid decorated with plumed rattlesnakes. The Ciudadela has the same 286° alignment found at the Upper Temple of the Jaguars. Located at the heart of Teotihuacan on the principal axis of the site, the pyramid's orientation coincides with the setting position of the Pleiades around A.D. 1 when the city plan was laid out (Dow 1967:328-329, Figs. 2, 6, 7). Teotihuacan may be the origin of this distinctive alignment, which may relate to observations of the Pleiades and Venus on the horizon just prior to the rainy season. Studying the pattern of the Pleiades conjunction with the evening star before the onset of the rainy season (Tuckerman 1964), we find it is a very ancient phenomenon, dating at least as far back as the founding of Teotihuacan.

Mexican legends say Quetzalcoatl invented the calendar. Aveni (1986) has pointed out a relationship between the 260 day calendar and the periods of visibility of the morning and evening star. Ancient observations of these periods may have influenced the development of the unique 260 day calendar. Venus was probably observed carefully in relation to the background of stars during the periods of visibility as morning and evening star. Observation of Venus in conjunction with the Pleiades dates back to the Early Classic period or earlier. Ancient priest-astronomers probably encoded a pattern linking Venus, the Pleiades and the rainy season in iconography and orientations of Venus temples throughout Mesoamerica.

NOTES

1. I would like to thank Anthony Aveni, Sylviane Bouchet, Ramon Carrasco, Clemency Coggins, Lawrence Desmond, Keff Karl Kowalski and Virginia Miller for reading and commenting on earlier drafts of this paper.

2. Aveni (1980:266; Table 10) provides two slightly different dates for the last appearance of the Pleiades around A.D. 1000 at the latitude of Chichen Itza. This difference of about a week (April 19 versus April 28) results from calculations based on varying assumptions and atmospheric conditions (Aveni, personal communication). Since precession slowly affects the disappearance dates as well, it is probably not possible to precisely determine the exact day of disappearance.

3. The inferior and superior conjunction phases of Venus are estimated based on Venus being in a celestial longitude of 7° or closer to the sun (as in Lounsbury 1982). These estimates, however, can only be considered to be approximate. As Šprajc (1988) notes, the exact date of first and last appearances of Venus before and after conjunction cannot be reliably reconstructed due a number of complex variables relating to the arcus visionis - the necessary depression of the sun below the horizon.

Table 1
Venus conjunction with the Pleiades
Based on Hawkins (1966) and Tuckerman (1964)

Pleiades position and Venus longitude approximate

Year	Pleiades position	Venus longitude	Venus phase	Julian date	
A.D.670	37	37	IC	April 24,	670
	37	37	MS	June 6,	671
	37	37	ES	March 25,	672
	37	37	MS	May 9,	673
	37	37	ES	March 20,	674
	37	37	MS	June 10,	674
	37	37	ES	April 9,	675
	37	37	MS	May 23,	676
	37	37	ES	March 12,	677
	37	37	IC	April 24,	678
A.D.900	40	40	MS	May 10,	900
	40	40	ES**	April 13,	901
	40	40	MS	June 11,	901
	40	40	ES	April 10,	902
	40	40	MS	May 26,	903
	40	40	ES	March 13,	904
	40	40	SC	April 25,	905
	40	40	MS	June 7,	906
	40	40	ES	March 27,	907
	40	40	MS	May 10,	908
A.D.1500	49	49	ES	April 7,	1500
	49	49	MS	May 22,	1501
	49	49	ES	March 16,	1502
	49	49	ES*	April 23,	1503
	49	49	MS	June 5,	1504
	49	49	ES	March 24,	1505
	49	49	MS+	May 9,	1506
	49	49	MS	June 17,	1507
	49	49	ES	April 7,	1508

MS=morning star
ES=evening star
SC=superior conjunction
IC=inferior conjunction
*=first appearance ES after invisibility in superior conjunction
**=last appearance ES before invisibility in inferior conjunction
++=first appearance MS after invisibility in inferior conjunction
+=last appearance MS before invisibility in superior conjunction

Table 2
Venus in conjunction with the Pleiades
A.D. 869-889

Dates in Julian calendar, add 4 days for Gregorian dates
Pleiades position and Venus longitude approximate

Pleiades with Venus	Pleiades	Venus	Venus phase	first of ES	last of ES
5/9/869	40	40	MS	---------	4/25/669
6/5/869	40	40	MS	same as above	
4/12/870	40	40	ES	3/26/870	11/28/870
5/27/871	40	40	MS	10/22/871	--------
3/15/872	40	40	ES	---------	7/5/872
4/27/873	40	40	SC	5/29/873	--------
6/8/874	40	40	MS	---------	2/10/874
3/29/875	40	40	ES	1/9/875	9/16/875
5/12/876	40	40	MS	8/6/876	--------
3/13/877	40	40	ES	---------	4/23/877
5/2/877	40	40	IC	same as above	
6/8/877	40	40	MS	same as above	
4/12/878	40	40	ES	3/24/878	11/26/878
5/27/879	40	40	MS	10/20/879	--------
3/15/880	40	40	ES	---------	7/1/880
4/27/881	40	40	SC	5/27/881	--------
6/7/882	40	40	MS	---------	2/8/882
3/29/883	40	40	ES	1/7/883	9/9/883
5/12/884	40	40	MS	8/9/884	--------
4/26/885	40	40	IC	---------	4/21/885
5/10/885	40	40	MS	same as above	
4/12/886	40	40	ES	3/22/886	11/24/886
5/27/887	40	40	MS	10/18/887	--------
3/14/888	40	40	ES	---------	6/29/888
4/26/889	40	40	SC	5/30/889	--------

MS=morning star
ES=evening star
SC=superior conjunction
IC=inferior conjunction

Bibliography

Andrews, E. W IV and E. W. Andrews V
1980 Excavations at Dzibilchaltun, Yucatan, Mexico. Tulane University, Middle American Research Institute, Publication 48.

Aveni, A. F.
1971 Tables of rise-set azimuths and heliacal rise-set dates. Colgate University Computer Center, Hamilton, New York.

1972 Astronomical tables intended for use in astro-archaeological studies. American Antiquity 37(4):531-540.

1980 Skywatchers of ancient Mexico. University of Texas Press, Austin.

1986 The real Venus-Kukulcan in the Maya inscriptions and alignments. Paper presented at the Sixth Mesa Redonda de Palenque, Mexico.

Aveni, A. F. and S. L. Gibbs
1976 On the orientation of Precolumbian buildings in Central Mexico. American Antiquity 41(4):510-517.

Aveni, A. F., S. L. Gibbs and H. Hartung
1975 The Caracol Tower at Chichen Itza: An ancient astronomical observatory? Science 188 (4192):977-985.

Closs, M. P., Aveni, A. F. and Crowley, B.
1984 The planet Venus and Temple 22 at Copan. Indiana 9(1):221-247.

Cohodas, M.
1978 The Great Ball Court at Chichen Itza, Yucatan, Mexico. Garland Publishing Co., New York.

Coggins, C.
1983 The Stucco decoration and architectural assemblage of Str. 1-sub, Dzibilchaltun, Yucatan, Mexico. Tulane University, Middle American Research Institute, Publication 49.

1984 "Murals in the Upper Temple of the Jaguars, Chichen Itza." in Cenote of Sacrifice: Maya treasures of the Sacred Well at Chichen Itza. edited by C. Coggins and O. C. Shane, pp. 157-165. University of Texas Press, Austin.

1986 A New Sun at Chichen Itza. Paper presented at the Second Oxford International Conference on Archaeoastronomy. January 1986, Merida, Yucatan, Mexico, submitted for publication.

Dow, J. W.
1967 Astronomical orientations at Teotihuacan, a case study in astro-archaeology. American Antiquity 32(3):326-334.

Hawkins, G. S.
1966 Astro-archaeology. Research in Space Science, Special Report, 226. Smithsonian Institution Astrophysical Observatory, Cambridge, Mass.

Hassig, R.
1988 Structure and development of the Aztec empire. Paper presented at the Fifty-third Annual Meeting of the Society for American Archaeology, Phoenix, Arizona.

Kelley, D. H.
1982 "Notes on Puuc inscriptions and history." Supplement to The Puuc: New Perpectives, edited by L. Mills. Central College, Pella, Iowa.

Klein, C.
1976 The identity of the central deity of the Aztec Calendar Stone. Art Bulletin LVIII(1):1-12

Kowalski, J. K.
1987 The House of the Governor. University of Oklahoma Press, Norman and London.

Krochock, R. J.
1988 The Hieroglyphic Inscriptions and Iconography of Temple of the Four Lintels and Related Monuments, Chichen Itza, Yucatan, Mexico. Masters of Arts Thesis, University of Texas, Austin.

Krupp, E. C.
1982 The "binding of the years," the Pleiades, and the nadir sun. Archaeoastronomy V(1):9-13.

1983 Echos from Ancient Skies. Random House, New York.

Kubler, G.
1961 Chichen Itza y Tula. Estudios de Cultura Maya 1:47-80.

Lincoln, C.
1986 The chronology of Chichen Itza: A review of the literature. In Late Lowland Maya Civilization: From Classic to Postclassic, edited by E.W. Andrews V and J. A. Sabloff, pp. 141-196. Santa Fe: School of American Research. Advanced Seminar Series.

1988 Dual Kingship at Chichen Itza, Yucatan. Paper presented at the Fifty-third Annual Meeting of the Society for American Archaeology, Phoenix, Arizona.

Lounsbury, F.
1982 Astronomical knowledge and its uses at Bonampak, Mexico. In Archaeoastronomy in the New World, edited by A. F. Aveni, pp. 143-166. Cambridge University Press, Cambridge:

Marquina, I.
1964 Arquitectura Prehispanica. INAH, Mexico.

Maudslay, A.P.
1889-1902 Archaeology. Biologia Centrali-Americana (56-60). Vol. 3. Porter, London.

Milbrath, S.
1980 Star gods and astronomy of the Aztecs. In La antropologia americanista en la actualidad, Homenaje a Raphael Girard, Vol. 1. pp. 289-305. Editores Mexicanos Unidos, Mexico.

1981 Astronomical imagery in the serpent sequence of the Codex Madrid. In Archaeoastronomy in the Americas, edited by R. A. Williamson, pp. 263-284. Ballena Press, Los Altos and Center for Archaeoastronomy, College Park.

1983 The "sun king" in ancient America. Paper presented at the Annual Meeting of the College Art Association, Philadelphia, 1983.

Miller, A. G.
1977 Captains of the Itza: Unpublished Mural Evidence from Chichen Itza. In Social Process in Maya Prehistory: Studies in Honour of Sir Eric Thompson, edited by N. Hammond, pp. 197-225. Academic Press, New York.

Morris, E. H., Charlot, J. and A. A. Morris
1931 The Temple of the Warriors. Carneige Institution of Washington, Publication 406. 2 Vols.

Parsons, L. A.
1969 Bilbao, Guatemala, Vol. 2. Publications in Anthropology, no. 12. Milwaukee Public Museum, Milwaukee.

Pasztory, E.
1972 The historical and religous significance of the Middle Classic ball game. In Religion en Mesoamerica. XII Mesa Redonda, pp. 441-455. Sociedad Mexicana de Antropologia, Mexico.

Perez Toro, A.
1942 La Milpa. Merida, Yucatan, Mexico.

Pina Chan, R.
1977 Quetzalcoatl: Serpiente Emplumada. Fondo de Cultura Economica. Mexico.

Reed, N.
1964 The Cast War of Yucatan. Stanford University Press, Stanford.

Ruppert, K.
1931 The Temple of the Wall Panels. Carnegie Institution of Washington, Publication 403, contribution 3, Washington.

Sahagun, B. de
1950-1982 The Florentine Codex. Translated by A. J. O. Anderson and C. E. Dibble. School for American Research, the University of Utah Press, Santa Fe, New Mexico. 13 Vols.

Schele, L. and M. Miller
1986 The Blood of Kings: Dynasty and Ritual in Maya Art. Kimbell Art Museum, Fort Worth.

Šprajc, Ivan
1988 Venus-rain-maize complex in Mesoamerica: associated with the evening star? ms.

Tozzer, A.
1941 Landa's Relacion de las cosas de Yucatan. Papers of the Peabody Museum, Harvard University, Vol. 18. Cambridge, Mass.

1957 Chichen Itza and its Cenote of Sacrifice. Memoirs of the Peabody Museum, Harvard University, Vols. 11-12. Cambridge, Mass.

Trujillo, C. A. E.
1946 Enciclopedia Yucatanense. Vol. VI. Gobierno de Yucatan, Mexico.

Vivo Escoto, J. A.
1964 Weather and climate of Mexico and Central America. In Handbook of Middle American Indians Vol. 1, edited by Robert C. West, pp. 187-215. University of Texas Press, Austin.

Wren, L.
1986a The Great Ball Court Stone from Chichen Itza. ms. Paper presented at the Mesa Redonda de Palenque, 1986.

1986b Elite interaction during the Terminal Classic period of the northern Maya Lowlands: Evidence from the reliefs of the North Temple of the Great Ball Court at Chichen Itza. ms.

THE RELATIONSHIP BETWEEN THE VENUS TABLE AND AN ALMANAC IN THE DRESDEN CODEX

Victoria R. Bricker
Tulane University

It has long been known that the Dresden Codex, a Precolumbian Maya hieroglyphic book, contains a table that correlates the movements of Venus with the haab of 365 days (Förstemann 1901). What has not been recognized until now is that one of the almanacs in the same Codex has a similar function, correlating the Venus synodical period with the tzolkin of 260 days. The relationship between the Venus table and the Venus almanac is both calendrical and iconographic. It is, therefore, essential to have an understanding of the general characteristics of the Venus table before examining the almanac to which it seems to be related.

The Venus table appears on pages 24 and 46 to 50 of the Dresden Codex. It consists of two parts: (1) a preface, on page 24 (Figure 1), with instructions for calculating a long-count date to be used for entering the table and a table of multiples for recycling it, and (2) the table proper on the remaining five pages (e.g., Figure 2). Each page of the table proper covers a period of 584 days, subdivided into smaller intervals of 236, 90, 250, and 8 days. 584 is the closest integral approximation of the mean length of a Venus synodical period, which is 583.92 days (Stahlman and Gingerich 1963:xv). Each row of the table correlates five Venus periods with eight haabs (= 2920 days in both cases). The thirteen rows in the table sum to 37,960 days (= 13 x 2920). This total is equivalent to two calendar rounds of 18,980 days and equates 104 haabs with 65 Venus revolutions (Lounsbury 1983; Thompson 1972:62-64).

The tzolkin dates mentioned in each row of each page of the table refer to the four stations of Venus: last appearance as a morning star, first and last appearance as an evening star, and first appearance as a morning star (e.g., Figure 2). Only the value representing the disappearance interval before and after inferior conjunction corresponds to the true mean (8 days) (Aveni 1983); the others are canonical values for visibility as a morning star (236 days), invisibility before and after superior conjunction (90 days), and visibility as an evening star (250 days). The true means for these periods are 263 days, 50 days, and 263 days, respectively (Aveni 1980:89).

Sky bands are prominently displayed on four of the five pages of the Venus table. The uppermost picture on the first page of the table depicts a figure seated cross-legged on a double sky band containing a Sun glyph and two so-called star glyphs (Figure 2a). Floyd Lounsbury's (1983) date for entering that table, 10.5.6.4.0 1 Ahau 18 Kayab (23 November A.D. 934), leads to a date in row 1, column 3 of that page correlating several events of interest to the ancient Maya. The date in question is 10.5.7.14.16 5 Cib 4 Yax. It corresponds to 21 June A.D. 936 in the Gregorian calendar, which is the date of a summer solstice (Kluepfel 1986). It is also a date of Venus' last visibility as an evening star (Table 1). It is worth noting that the astronomical events coincide only if one relates the Maya and Western calendars by using the 584,283 correlation constant, which is the one that agrees best with historical sources; the 584,285 constant preferred by Lounsbury (1983:5-6) on the basis of a misinterpretation of astronomical data (Aveni 1986) yields a date for the relevant row and column of the Venus table that is two days after the summer solstice and after Venus has ceased to be visible as an evening star. The question of the visibility of Venus in late June of A.D. 936 is addressed by comparing the values in the rightmost column of Table 1 with the so-called arcus visionus for Venus at "elast," which may be defined in the present instance as the difference in altitude between the Sun (below the horizon) and Venus (on the horizon) at dusk on the last day of visibility of Venus as an evening star (Neugebauer 1929:Bd. I, 153). The arcus visionus for Venus at elast, although subject to some observational variation, has been determined to be about 5.2° (Schoch 1924:734; Neugebauer 1938:313), which is just the angular distance shown in Table 1 for 21 June A.D. 936. By 22 June, and surely by 23 June (the date arrived at using the 584,285 correlation), Venus had moved so close to the Sun, approaching inferior conjunction, that it would not have been visible at dusk, being lost in the glow of the setting Sun in the brief time before it too moved below the horizon (Table 1).

The sky band picture on this page resembles a picture in the lower register of the seasonal table in the same Codex (Figure 3), which is also associated with a day of summer solstice (V. Bricker and H. Bricker 1988). The rain god Chac is shown seated on the skyband in the seasonal table; the figure sitting on the sky band in the corresponding picture of the Venus table has not been identified (Thompson 1972:67). The dates referring to these pictures suggest that they are probably conventionalized representations of the summer solstice.

The sky band in the picture from the Venus table contains two star glyphs and a kin or Sun glyph (under the left foot of the seated figure) (Figure 2). The glyph for Venus is also a star glyph, and because Venus is near the Sun, I assume that the star glyph closest to the kin glyph in the upper half of the sky band (beside the right foot) was intended to represent Venus. Mars and Jupiter were also visible on the night in question, and they were in near conjunction in the same quadrant of the sky as Venus (Table 2). I assume that the second star glyph, in the lower half of the sky band, refers to one or both of these other planets (Figure 2).

The last event that would have interested the ancient Maya is the middle of the haab, which occurred four days earlier on 1 Eb 0 Yax. The head of God K, who is shown as the patron for Eb years on the first page of the New Year's almanac on pages 25 to 28, juts out from the left side of the sky band (compare Figure 2a with Figure 4a). God K also appears in the picture at the bottom of the same page as the target of a dart hurled by the figure in the middle picture (Figure 2b and c). He is replaced by a jaguar in the corresponding picture on the next page of the table (Figure 5c). The second page of the New Year's almanac depicts a jaguar as the patron of Caban years (compare Figures 6a and 5c), and A.D. 937, one of the years referred to on the second page of the Venus table, was a Caban year. The lower picture on the third page of the Venus table represents the maize god as the victim of the dart (Figure 7c); the maize god is shown as the patron of Ik years on the third page of the New Year's almanac (compare Figures 8a and 7c), and A.D. 938, one of the years implied by the third page of the Venus table, was an Ik year. There is, then, a direct relationship between the order of pictures in the New Year's almanac on pages 25 to 28 and the order of pictures at the bottom of the Venus table on pages 46 to 48. The yearbearer is the victim of the Venus warrior shown in the middle picture on each page.

The first page of the Venus table lists thirteen canonical dates for the last appearance of Venus as an evening star before inferior conjunction. They appear in the third column on that page, and they are all Cib dates, but with different coefficients. For example, the canonical elast date in the first row is a day 5 Cib, the date in the row below it is a day 13 Cib, the date in the third row of that column is a day 8 Cib, and so on through the thirteen rows of the table (Table 3). Of these, only the first date coincides with a summer solstice, and it is the only date that is actually a date of last visibility of Venus as an evening star. The canonical length of the Venus period used in the table (584 days) is 0.08 days longer than its mean (583.92 days). This means that the canonical dates for elast slowly move ahead of the actual station, as is shown by the upper curve in Figure 9a. After A.D. 936, these dates fall in the period when Venus is not visible. The lower curve in Figure 9a graphs the relationship between the dates of summer solstice and the angular distance between Venus and the Sun. It shows that the Venus period recesses much more rapidly through the tropical year than it does through the Venus table. The recession is a result of the fact that eight tropical years of 365.2422 days is 2.34 days longer than five Venus periods of 583.92 days.

The fourth station mentioned on the first page of the table is the date of first appearance of Venus as a morning star after inferior conjunction. Thirteen canonical dates are given for this station, one in each row. They are all Kan dates, beginning with 13 Kan (Table 3). The arcus visionus for Venus at "mfirst" is about 5.7° (Schoch 1924:734; Neugebauer 1938:313). The canonical value for the period of invisibility at inferior conjunction given in the table is eight days. Although this value corresponds to the mean, the actual length of the period of invisibility can vary widely, from three to sixteen days (Aveni 1983). Because the interval between elast and mfirst was about eleven days in A.D. 936, the 13 Kan date mentioned for heliacal rise in the first row of the table fell about three days too early. In fact, it was not until the sixth row of the table that one of the canonical Kan dates coincided with the empirical date for mfirst.

Figure 9b expresses the relationship between canonical dates of mfirst (solid line) and dates of summer solstice (dotted line) in terms of the angular distance between Venus and the Sun on those dates. In this case, the two lines intersect in the fifth row of the table, instead of the first. The year of convergence was A.D. 968.

The Venus table begins by correlating a date of heliacal rise after inferior conjunction with the tzolkin day, 1 Ahau. Two calendar rounds (= 104 years) must pass before 1 Ahau can coincide with the same station again. This is because the 584-day canonical value for the Venus period employed by the table is not an even multiple of the 260-day tzolkin, or of one of its factors, 13 or 20. The table emphasizes the relationship between Venus periods and the *haab* at the expense of the *tzolkin*.

The actual length of the Venus period varies between 579.6 and 588.1 days (Aveni 1980:324, n12). For some purposes, as in the Venus table, the ancient Maya found it convenient to use the 584-day value that approximates the mean. But for other purposes, a slightly higher value, 585 days, which also falls within the range of variation, was more useful because it is divisible by 13, and it is possible to correlate it with the *tzolkin* cycle in less than seven years.

Most of the almanacs in the Maya codices have a 260-day structure. A few almanacs contain multiples of 260 days, and in every case they refer to some astronomical or seasonal phenomenon. For example, the almanac in the middle of pages 38 to 41 of the Dresden Codex covers a double-*tzolkin* period of 520 days, and the text and the pictures refer to eclipses and agricultural practices (V. Bricker and H. Bricker 1986). Similarly, the almanac on pages 22 and 23 of the Paris Codex is composed of seven *tzolkin* periods, and it contains pictures of the thirteen Maya zodiacal signs (Kelley 1976:45-50).

The almanac in the bottom register of pages 30 to 33 of the Dresden Codex covers a period of 2340 days, which equates four Venus periods of 585 days with nine *tzolkin* cycles of 260 days (Figure 10). The preface to the almanac consists of four columns and five rows of glyphs. Each column has a bar-and-dot numeral at the top, which functions as the coefficient of the day glyphs below it. Unlike the Venus table, the preface has no long-count date or table of multiples that might be used to correlate the almanac with the Christian calendar or recycle it. The abbreviated preface is what distinguishes almanacs from tables in the Maya codices.

The almanac proper is also divided into five rows of 117 days each. The average synodical period of Mercury is 115.88 days (Stahlman and Gingerich 1963:xv). The closest integral approximation of the mean is 116 days. A slightly larger value of 117 days for this period would have permitted the Maya to equate five Mercury revolutions with one Venus revolution and twenty Mercury revolutions with nine *tzolkin* periods. In other words, they could have commensurated these three cycles simply by adding one day to the length of the Venus and Mercury periods.

Each row of the almanac is further subdivided into nine thirteen-day intervals. Mercury, like Venus, has four stations, first and last visibility as a morning star and first and last visibility as an evening star. The mean disappearance interval before and after superior conjunction is 35 days; it is invisible for approximately five days around inferior conjunction. The periods of visibility as a morning or an evening star both average 38 days; according to Aveni (1980:89), the actual values fluctuate widely around this mean. It is interesting that the 38-day period when Mercury is visible as a morning or as an evening star is very close to three intervals of thirteen days (3 x 13 = 39).

The almanac contains nine pictures and captions (Figure 10). Two sets of bar-and-dot numbers lie between the pictures and their captions. The first set, in every case consisting of two bars and three dots in solid black, represent the thirteen-day intervals mentioned above. The numbers next to them, which are outlined in black, represent the coefficients of the tzolkin dates reached by adding thirteen days to the previous tzolkin date. Each column in the preface contains five day glyphs, which are the beginning dates for each row. Five rows sum to 585 days (5 x 117). There are four columns, so that the total length of the almanac is 2340 days, which is equal to nine tzolkin cycles or four Venus periods of 585 days and twenty Mercury periods of 117 days.

Seven of the nine pictures in the almanac depict the rain god Chac sitting in a tree. The caption above the first picture says that the tree is chac te (literally, 'red tree'; misspelled as chac e-te). This could be a reference to Neea fagifolia Heimerl., which has the same name in Modern Yucatec (Sosa, et al. 1985:146). The second tree is described as sac tun (literally, 'white stone'). Its identity is unknown. The third tree is called ek te (literally, 'black tree'). It may be a reference to the logwood tree (Haematoxylon campechianum L.), which is called ek in Modern Yucatec (Sosa, et al. 1985:117). The fourth tree is referred to as kan te (literally, 'yellow tree'). This is the name for Caesalpinia violacea (Miller) Standley in Modern Yucatec (Sosa 1985:108). The fifth tree appears in the third picture on page 32c. It looks like an agave, but I am unable to decipher the glyph that refers to it. The name of the sixth tree is dziu te. The Modern Yucatec cognate (dziu che) refers to Pithecellobium dulce (Roxb.) Benth. And the seventh tree is called chicah, which is the Mopan name for Bursera simaruba (L.) Sarg. (the Modern Yucatec cognate is chacah [Sosa, et al. 1985:39]).

The first four trees are associated with the four directions: east (lakin), north (xaman), west (chikin), and south (nohol). The other three trees do not have directional associations.

The middle picture on page 32c depicts the rain god Chac sitting on a sky band. It is similar to a picture on the first page of the Venus table, which I have shown refers to the summer solstice. Here, as on the first page of the Venus table, the sky band contains a star glyph. The picture in the almanac seems to be telling us that one of the tzolkin dates associated with it must correspond to the date of a summer solstice when Venus or some other planet was visible.

Each picture in the almanac is associated with twenty possible tzolkin dates, corresponding to the twenty rows that are implied by the four columns of day glyphs in its preface. One of the tzolkin dates that refers explicitly to the solstitial picture on page 32c is identical to the canonical elast date mentioned in the fifth row of the first page of the Venus table (Table 3). That date is 11 Cib, and it appears in the fourth row of the almanac at hand (Figure 10). This calendrical link between the almanac and the Venus table makes it possible to correlate the almanac with the Western calendar and to explore its astronomical implications in real time.

The 11 Cib date in the fifth row of the Venus table corresponds to 13 June A.D. 968. Venus was last visible as an evening star on the previous day (Table 4b). Five days earlier, Mercury had made its last appearance before inferior conjunction, at which time it was in conjunction with Venus at sunset (Table 4a). The summer solstice took place on 21 June, one day before Venus reappeared as a morning star (compare Table 4c and d). The solstitial picture in the almanac suggests that the Maya astronomer had expected the solstice to coincide with the first appearance of Venus as a morning star (he was only one day off). Mercury re-emerged as a morning star six days later (Table 4e). By then, it was less than one degree from Venus, and they would have looked like a double star.

The interval associated with the solstitial picture is thirteen days. Both the summer solstice and Venus' first pre-dawn appearance after inferior conjunction occurred during that interval. The Venus table lists 6 Kan as the canonical date for mfirst in that year (the actual date was one day later, 7 Chicchan). If the prediction had been accurate, the summer solstice would have coincided with mfirst. The picture in the almanac is faithful to the canonical date for mfirst that is given in the Venus table. The interest in the year A.D. 968 may be a result of the fact that it is the only year during the time period covered by the Venus table when mfirst could have coincided with the summer solstice (Figure 9b).

The solstitial picture is only relevant for the fourth row of the almanac. In other rows, however, the dates associated with it target Mercury's first appearance as a morning star (the arcus visionus for Mercury at mfirst is about 13° [Neugebauer 1938:313]). The relationship between the solstitial picture and Mercury's first appearance as a morning star in each row of the almanac is shown in Figure 11. The points on the x-axis refer to the dates in the intervals that are closest to mfirst. The 13° value for this station has been subtracted from the angular difference in the altitudes of Mercury and the Sun on those dates, and the results are plotted on the y-axis. The graph shows that the almanac successfully predicted mfirst in half the rows and that Mercury was a morning star on eighteen of the twenty days associated with the solstitial picture. This suggests that the primary purpose of the almanac was to chart the movements of Mercury, with mfirst centered on the solstitial picture and elast on the picture that precedes it. The periods of invisibility at superior conjunction and visibility as morning star and evening star would have been divided into modules of thirteen days and assigned to the other pictures.

Although the 117-day length of the rows in the almanac exceeded the mean length of the Mercury period by more than one day, the thirteen-day intervals into which the synodical period was divided made it possible to correlate one of the pictures with mfirst about half the time. The almanac is not a perfect Mercury calendar; rather, it represents the best compromise for commensurating Mercury and Venus periods with the Maya tzolkin. This must have been the intention of the Maya astronomer who was responsible for the layout of the almanac. We are fortunate that he included sufficient information to permit us to cross-date it with the Venus table and from there with the Gregorian calendar.

TABLE 1. Positions of Venus and the Sun at the moment of Venus-set on 21, 22, and 23 June A.D. 936 (Gregorian) in the Yucatan Peninsula (20°30' north latitude, 88°30' west longitude). Astronomical data obtained from Kluepfel (1986).

a. Positions as of 6:56 P.M. on 21 June.

	Ecliptic Longitude	Ecliptic Latitude	Azimuth	Altitude	Diff.
Venus	97.16°	-2.90°	291.98°	-0.10°	5.18°
Sun	90.09°	0°	297.61°	- 5.28°	

b. Positions as of 6:49 P.M. on 22 June.

	Ecliptic Longitude	Ecliptic Latitude	Azimuth	Altitude	Diff.
Venus	96.57°	-3.13°	291.75°	-0.06°	3.72°
Sun	91.04°	0°	296.91°	- 3.78°	

c. Positions as of 6:42 P.M. on 23 June.

	Ecliptic Longitude	Ecliptic Latitude	Azimuth	Altitude	Diff.
Venus	95.97°	-3.35°	291.53°	-0.03°	2.25°
Sun	91.99°	0°	296.23°	-2.28°	

TABLE 2. Positions of Mars, Jupiter, and the Sun about one hour after sunset (7:30 p.m.) on 21 June A.D. 936 (Gregorian) in the Yucatan Peninsula (20°30' north latitude, 88°30' west longitude). Astronomical data obtained from Kluepfel (1986).

	Ecliptic Longitude	Ecliptic Latitude	Azimuth	Altitude	Diff.
Mars	144.77°	1.09°	273.31°	35.68°	5.56°
Jupiter	138.22°	1.02°	277.58°	30.12°	
Sun	90.11°	0.00°	301.19°	-12.22°	

TABLE 3. Canonical dates for Venus stations on page 46 of the Dresden Codex.

Row	mlast Cib	efirst Cimi	elast Cib	mfirst Kan
1	3	2	5	13
2	11	10	13	8
3	6	5	8	3
4	1	13	3	11
5	9	8	11	6
6	4	3	6	1
7	12	11	1	9
8	7	6	9	4
9	2	1	4	12
10	10	9	12	7
11	5	4	7	2
12	13	12	2	10
13	8	7	10	5

TABLE 4. Positions of Mercury, Venus, and the Sun at the moment of sunset on 7, 12, 21, 22, and 28 June A.D. 968 (Gregorian) in the Yucatan Peninsula (20°30' north latitude, 88°30' west longitude). Astronomical data obtained from Kluepfel (1986).

			Ecliptic Longitude	Ecliptic Latitude	Azimuth	Altitude
(a)	7 June	Mercury	89.54°	-2.38°	288.59°	11.29°
	6:27 p.m.	Venus	91.13°	-0.53°	289.92°	13.36°
		Sun	77.00°	0.00°	294.58°	0.00°
(b)	12 June	Mercury	87.27°	-3.63°	289.89°	3.73°
	6:30 p.m.	Venus	88.47°	-1.70°	291.32°	5.49°
		Sun	81.76°	0.00°	295.12°	-0.29°
(c)	21 June	Mercury	82.45°	-4.68°	71.83°	4.81°
	5:15 a.m.	Venus	83.28°	-3.57°	70.48°	4.49°
		Sun	89.81°	0.00°	64.66°	-0.15°
(d)	22 June	Mercury	82.02°	-4.68°	72.32°	6.08°
	5:15 a.m.	Venus	82.69°	-3.76°	71.20°	5.82°
		Sun	90.76°	0.00°	64.65°	-0.20°
(e)	28 June	Mercury	80.91°	-4.22°	74.12°	12.58°
	5:15 a.m.	Venus	79.71°	-4.71°	75.05°	13.52°
		Sun	96.48°	0.00°	64.68°	-0.53°

REFERENCES CITED

Aveni, Anthony F.

1980 <u>Skywatchers of Ancient Mexico</u>. University of Texas Press, Austin.

1983 The Moon and the Venus table: An example of commensuration in the Maya calendar. A paper presented at the conference on ethnoastronomy, Smithsonian Institution, Washington, D.C., September, 1983.

1986 The real Venus-Kukulcan in the Maya inscriptions and alignments. A paper presented at the Sixth Mesa Redonda de Palenque, June, 1986.

Bricker, Victoria R. and Harvey M. Bricker

1988 The seasonal table in the Dresden Codex and related almanacs. <u>Archaeoastronomy</u>, supplement to <u>Journal for the History of Astronomy</u> 12(<u>JHA</u> 19):S1-S60.

1986 Archaeoastronomical implications of an agricultural almanac in the Dresden codex. <u>Mexicon</u> 8:29-35.

Förstemann, Ernst

1901 <u>Commentar zur Mayahandschrift der Königlichen öffentlichen Bibliothek zu Dresden</u>. Verlag von Richard Bertling, Dresden.

Kelley, David H.

1976 <u>Deciphering the Maya script</u>. University of Texas Press, Austin and London.

Kluepfel, Charles

1986 <u>Planets</u> (an astronomical software package available from its author).

Lounsbury, Floyd G.

1983 The base of the Venus table of the Dresden Codex, and its significance for the calendar-correlation problem. In <u>Calendars in Mesoamerica and Peru: Native American computations of time</u>, edited by Anthony F. Aveni and Gordon Brotherston, pp. 1-26. BAR International Series 174, Oxford.

Neugebauer, P. V.

1929 <u>Astronomische Chronologie</u>. Walter de Gruyter & Co., Berlin.

1938 Tafeln zur Berechnung der jährlichen Auf- und Untergänge der Planeten. <u>Astronomische Nachrichten</u> 214(6331):313-322.

Schoch, Carl

1924 The 'arcus visionus' of the planets in the Babylonian observations. <u>Miscellaneous Notes of the Royal Astronomical Society</u> 84:731-734.

Sosa, Victoria, J. Salvador Flores, V. Rico-Gray, Rafael Lira, and J. J. Ortiz
1985 *Etnoflora yucatanense: Lista florística y sinonimia maya.* Instituto Nacional de Investigaciones sobre Recursos Bióticos, Xalapa.

Thompson, J. Eric S.
1972 *A commentary on the Dresden Codex.* Memoirs of the American Philosophical Society, 93, Philadelphia.

Villacorta C., J. Antonio and Carlos A. Villacorta
1976 *Códices mayas.* 2nd ed. Tipografía Nacional, Guatemala.

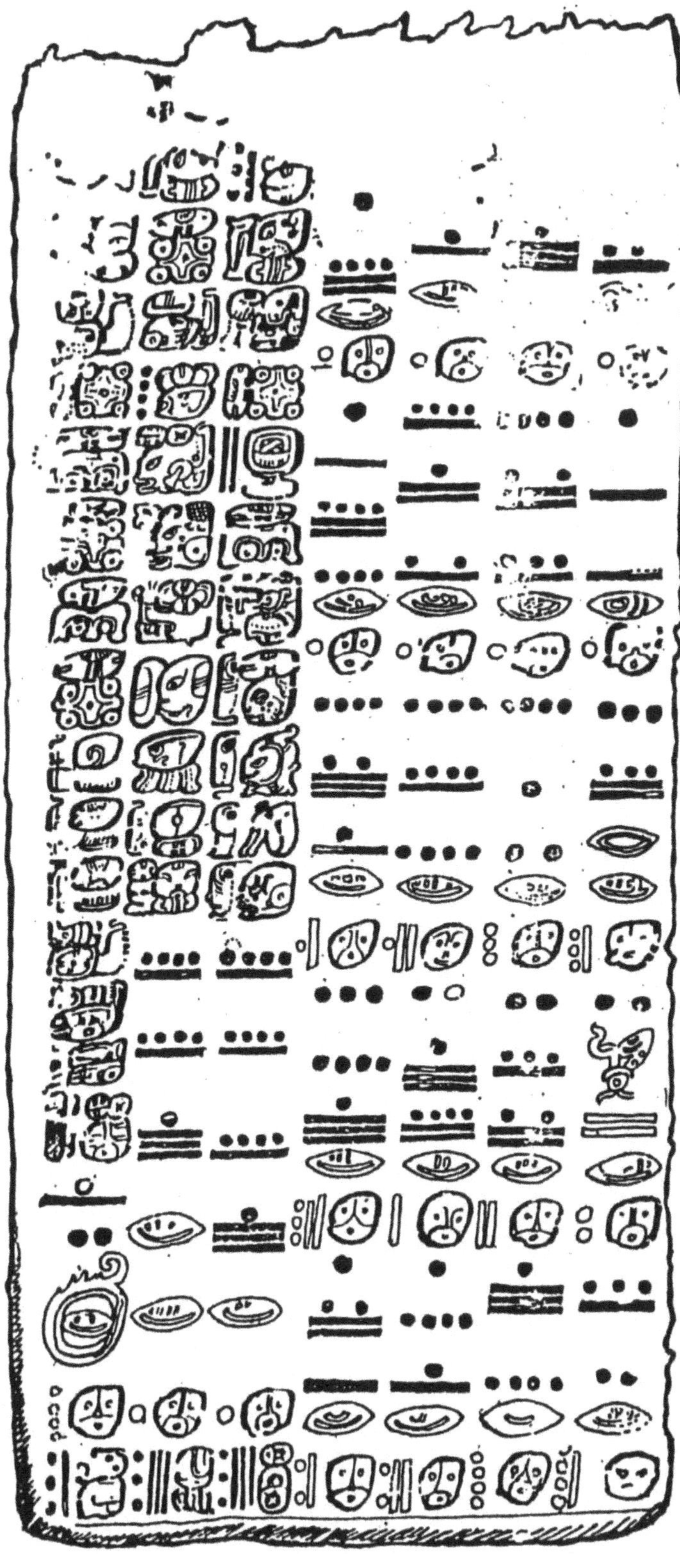

Figure 1 The preface to the Venus table on page 24 of the Dresden Codex (after Villacorta C. and Villacorta 1976:58).

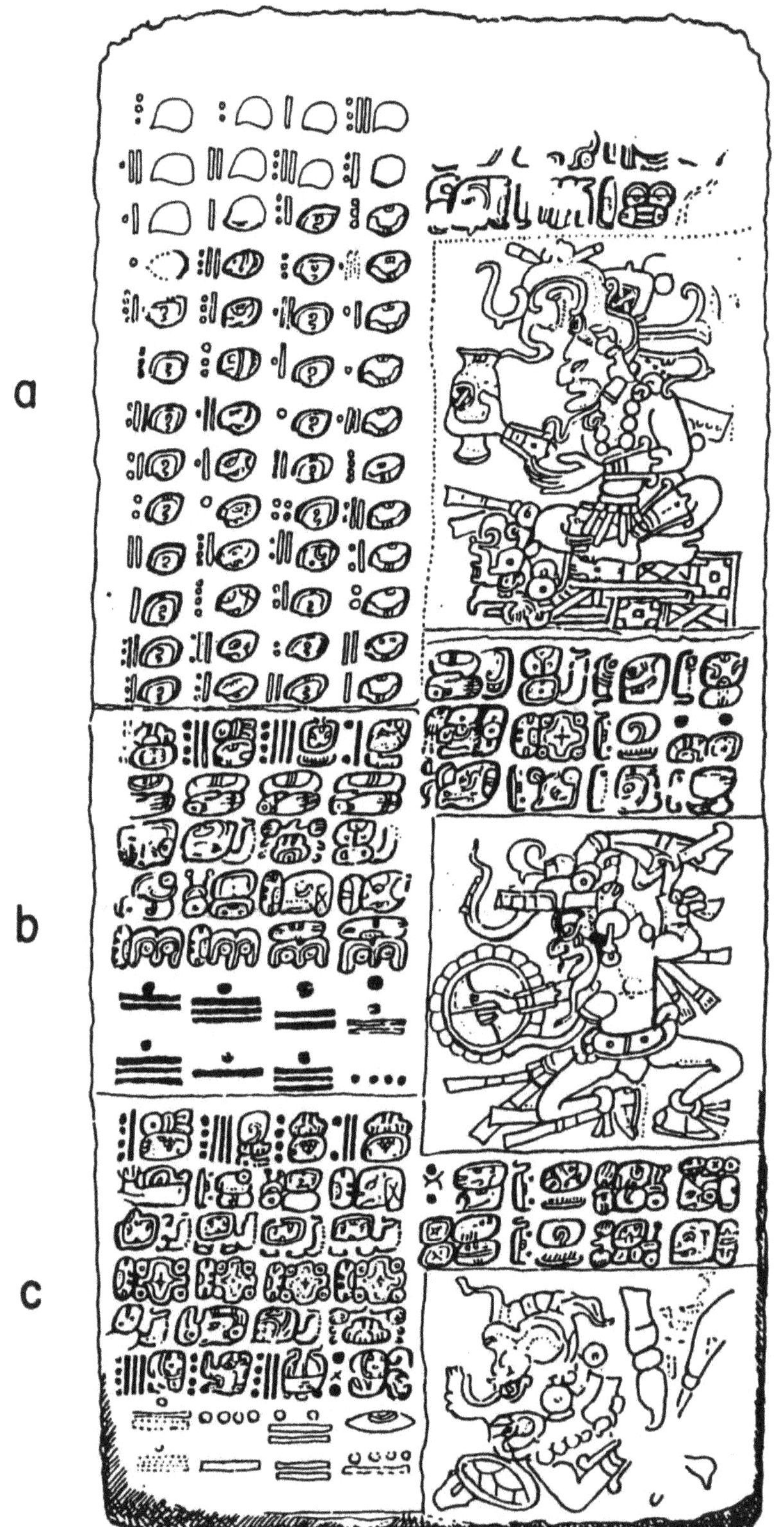

Figure 2 Page 46 of the Dresden Codex (after Villacorta C. and Villacorta 1976:102).

Figure 3 The summer solstice picture on page 66b of the Dresden Codex (after Villacorta C. and Villacorta 1976:142).

Figure 4 Page 25 of the Dresden Codex (Villacorta C. and Villacorta 1976:60).

a

b

c

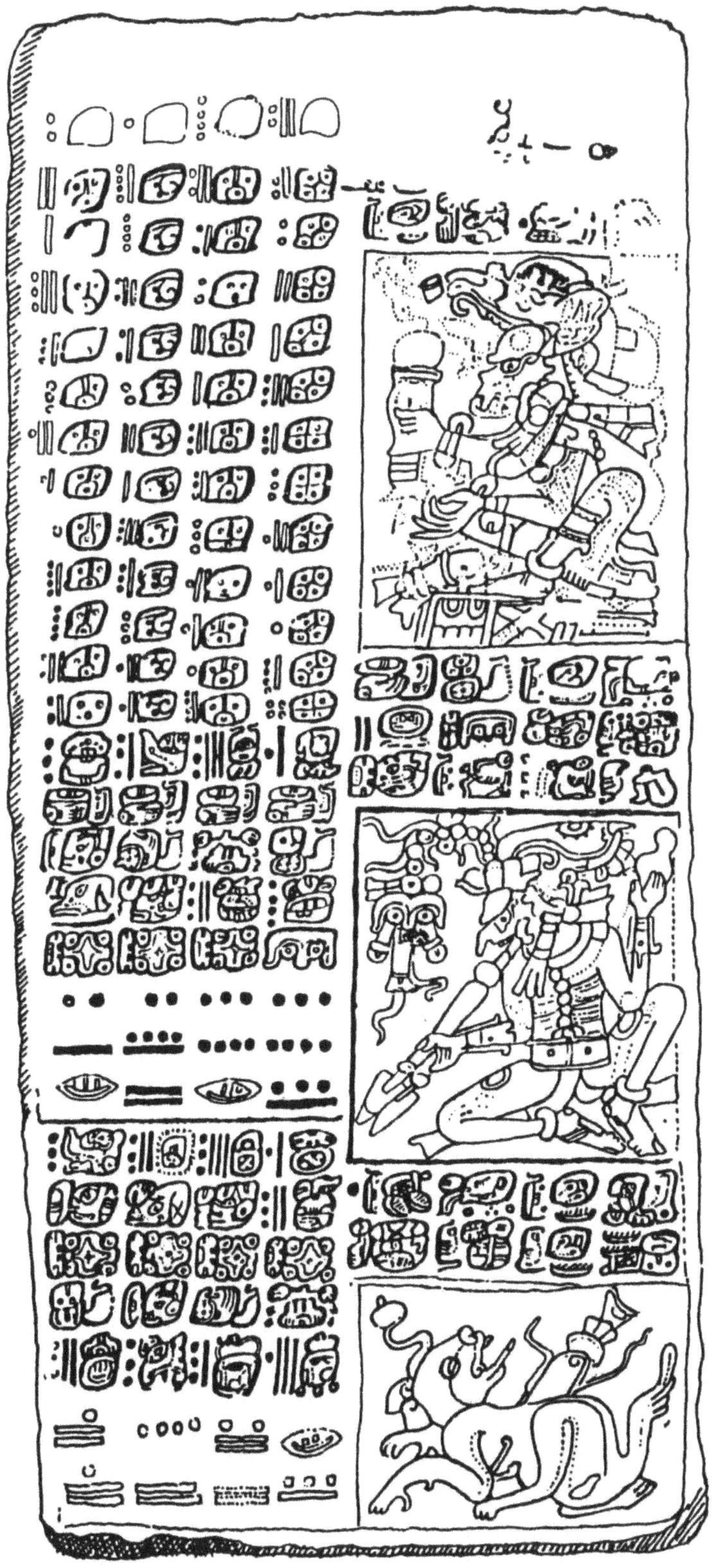

Figure 5 Page 47 of the Dresden Codex (after Villacorta C. and Villacorta 1976:104).

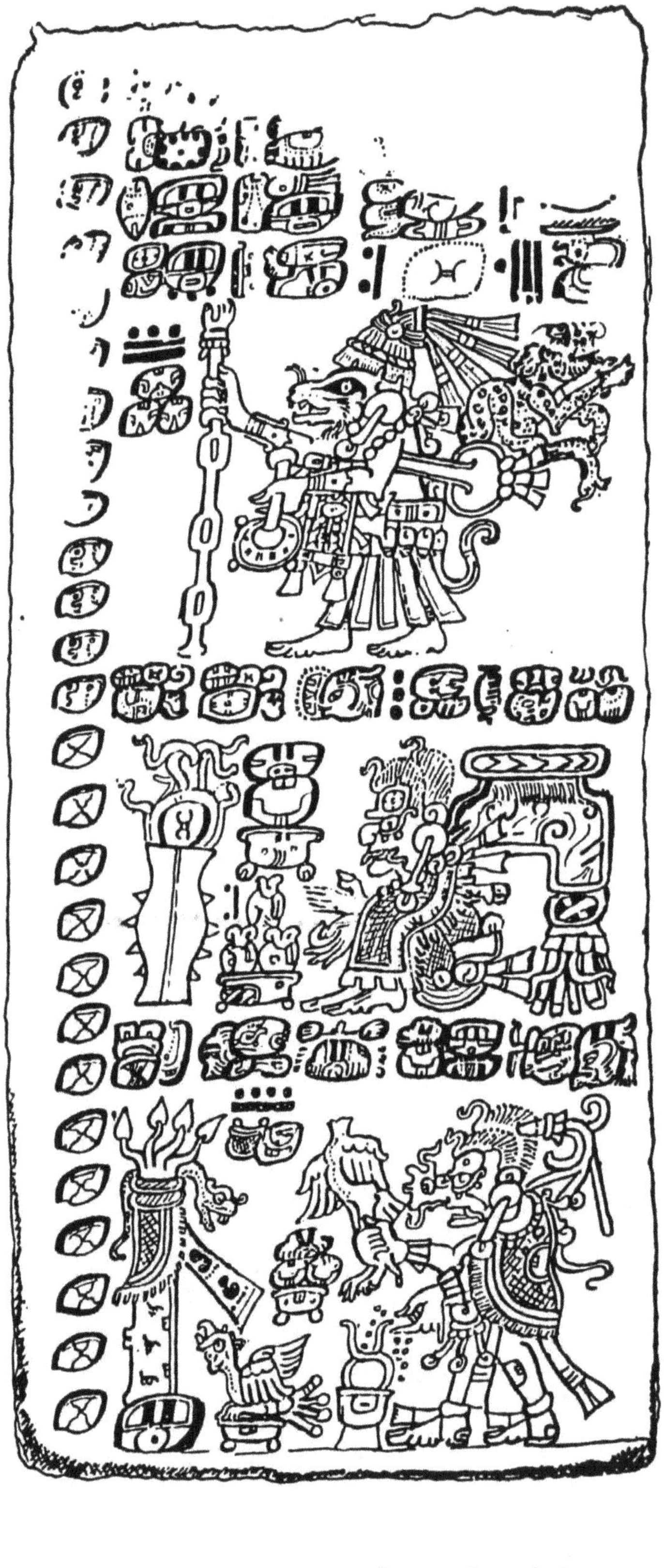

Figure 6 Page 26 of the Dresden Codex (after Villacorta C. and Villacorta 1976:62).

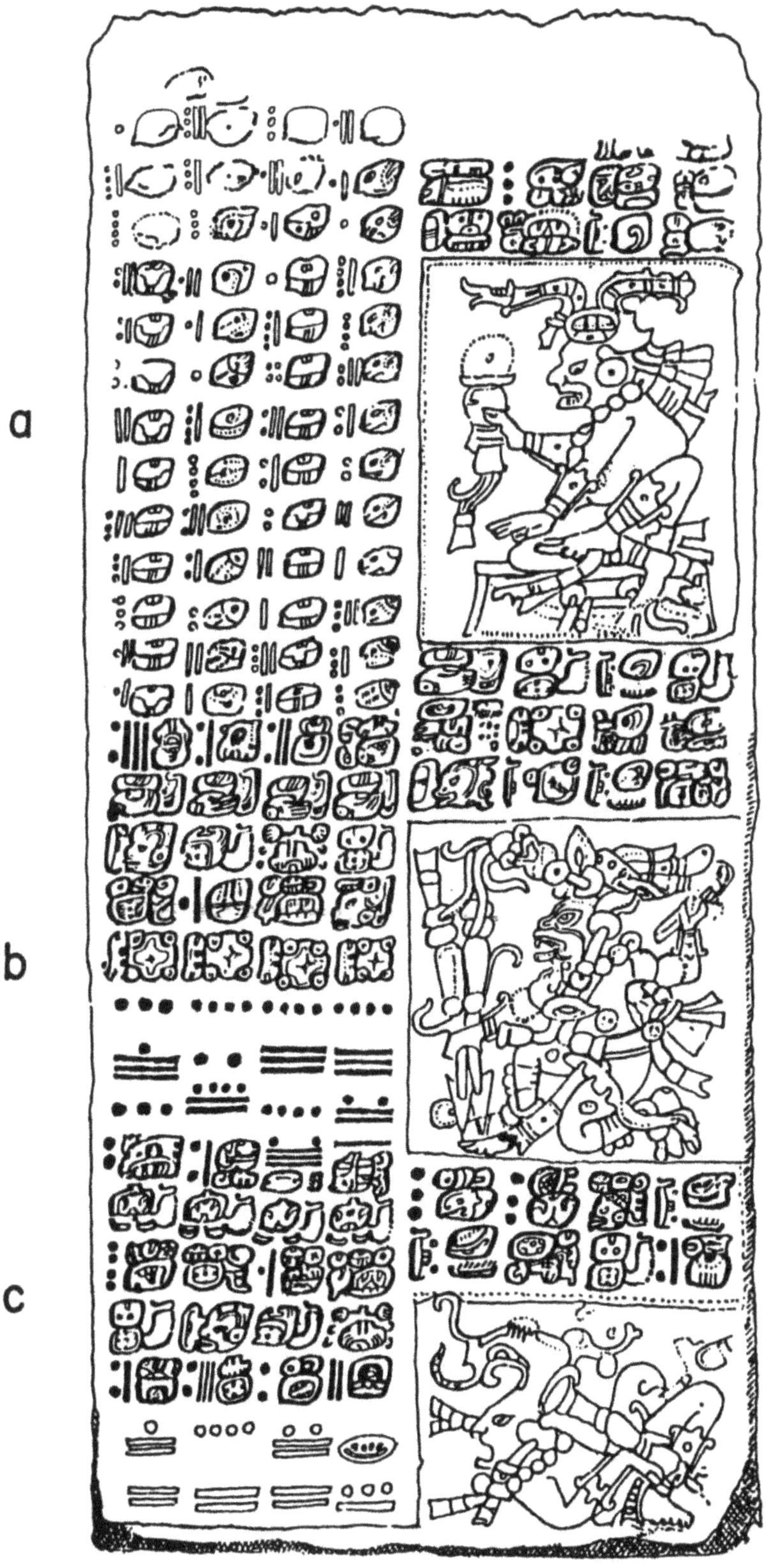

Figure 7 Page 48 of the Dresden Codex (after Villacorta C. and Villacorta 1976:106).

a

b

c

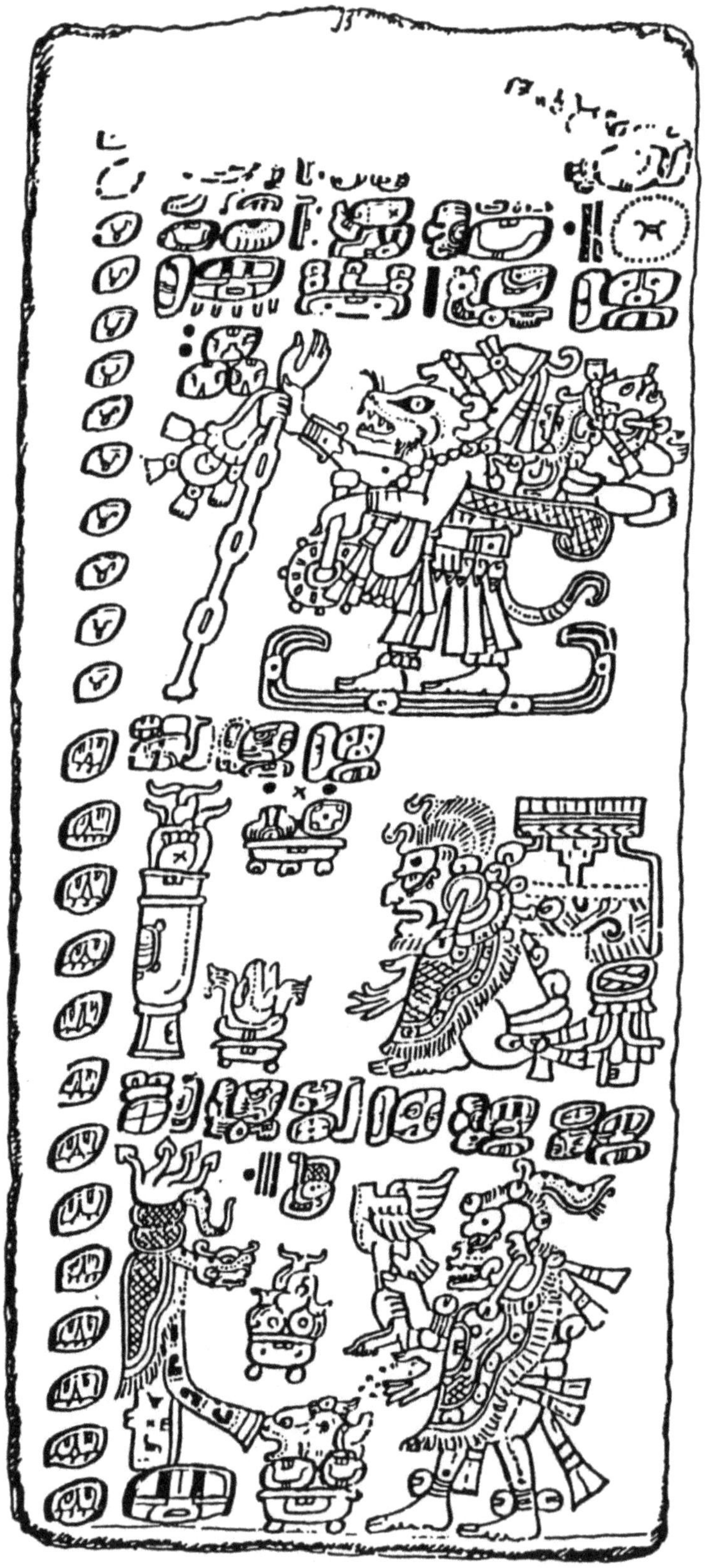

Figure 8 Page 27 of the Dresden Codex (after Villacorta C. and Villacorta 1976:64).

a

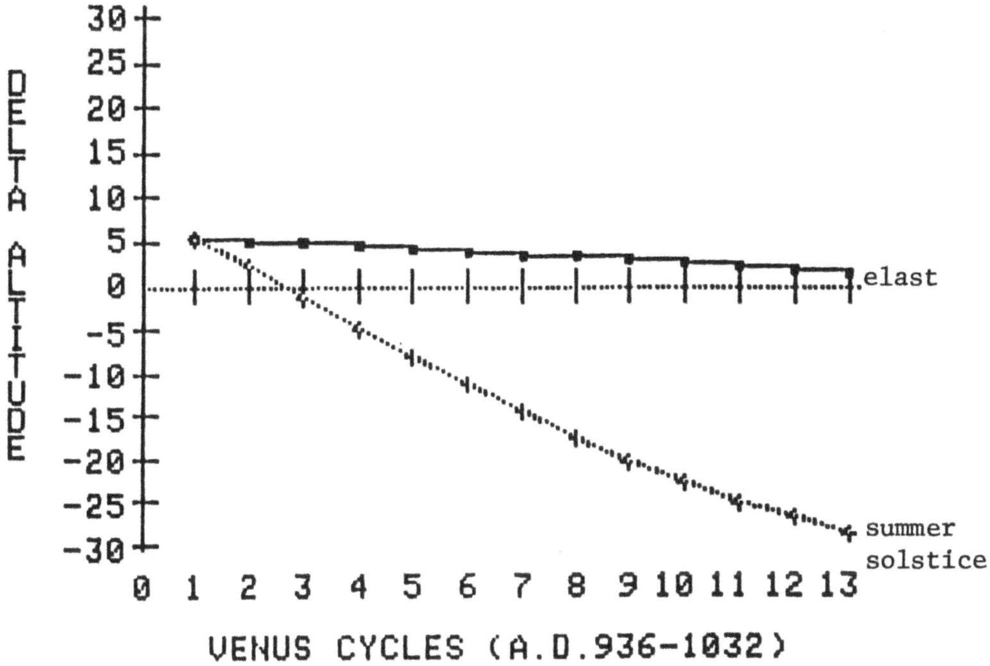

b

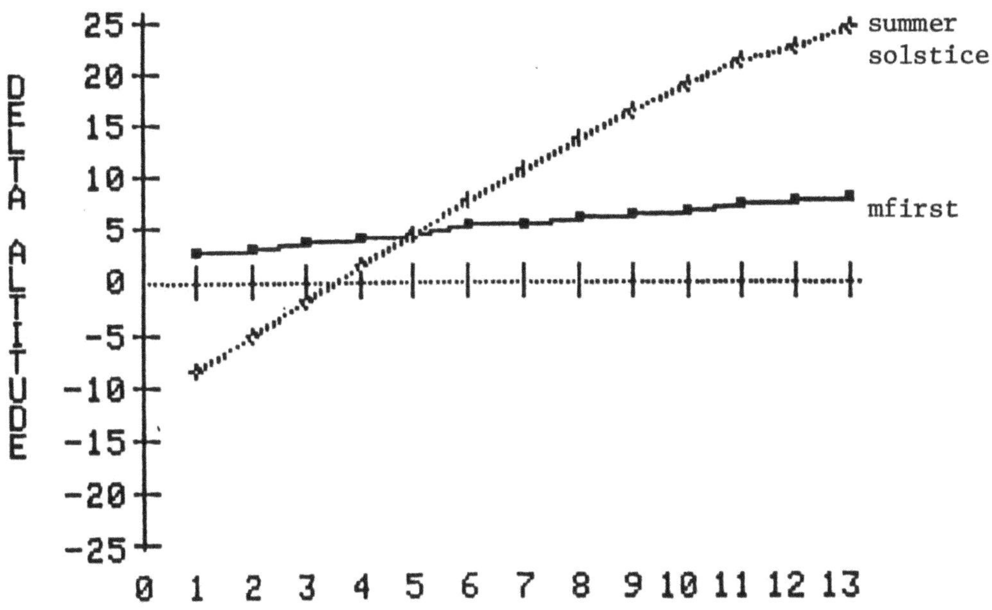

Figure 9 The relationship between the summer solstice and two Venus stations on page 46 of the Dresden Codex. (a) elast. (b) mfirst.

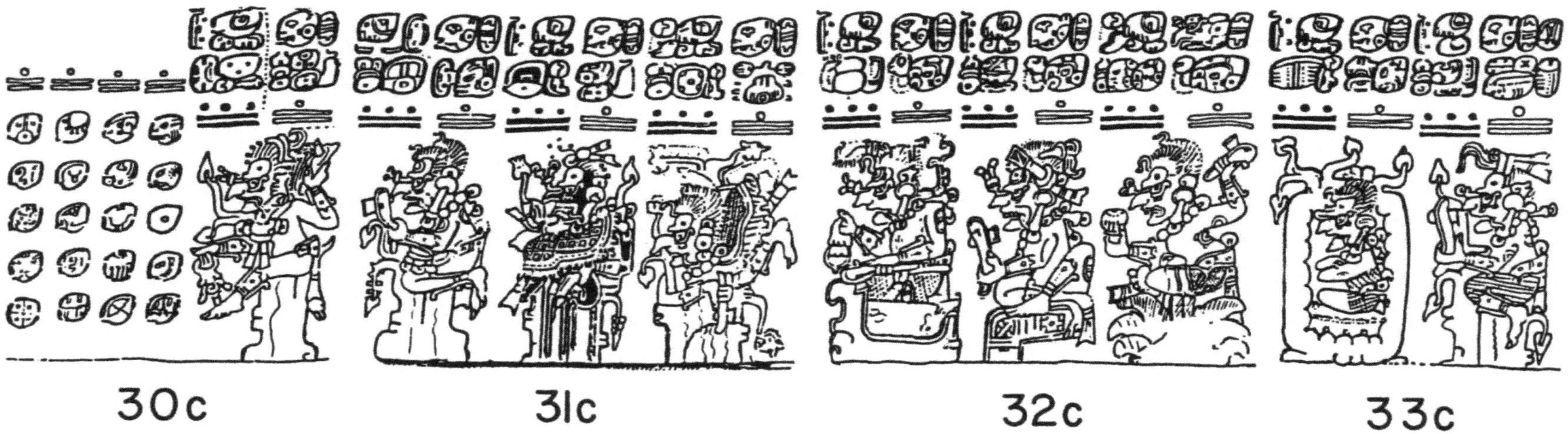

Figure 10 The almanac on pages 30c–33c of the Dresden Codex (after Villacorta C. and Villacorta 1976: 70, 72, 74, 76).

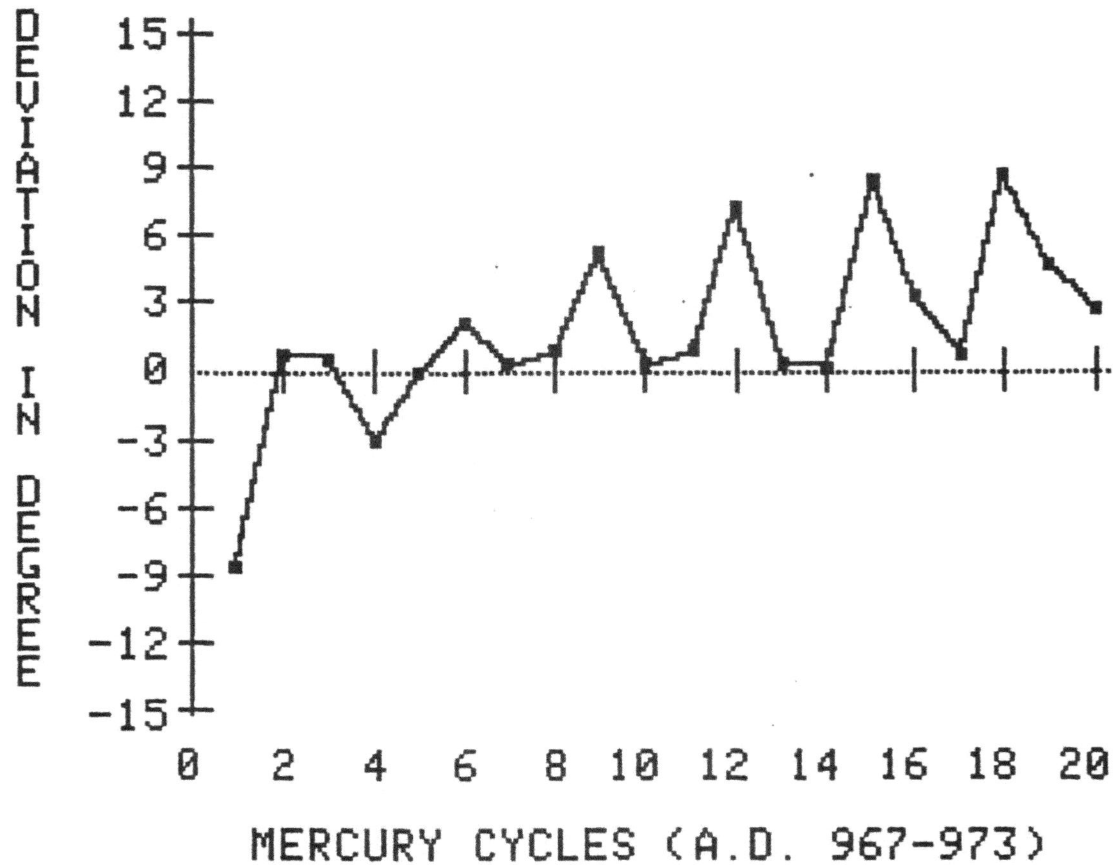

Figure 11 Deviation (in degrees) of the difference between the altitudes of Mercury and the Sun and the _arcus visionus_ for mfirst (13°) on dates associated with the solstitial picture in each row of the almanac on pages 30c-33c of the Dresden Codex.

FRANZ TICHY

Measurement of Angles in Mesoamerica. Necessity and Possibility.

1. Necessity of measurement of angles.

In the architecture of Mesoamerica both in the groundplans of buildings and in the settlement and field patterns several clearly distinguishable alignments have been discovered. Anthony Aveni and Horst Hartung have determined the directions, i.e. the bearings of many buildings, collected further data and laid the foundations for farther research (see Aveni 1980 appendix A; Aveni and Hartung 1986). The best known is the 15 1/2° E of N direction of the Teotihuacan grid in a clockwise direction. Others such as the axis of Cholula is skewed clockwise 25–26° and the pattern of the town and Templo Mayor of Tenochtitlan-Mexico with the 7° deviation E of N are also remarkable.

The great variety of alignments in the Maya region is confusing, it still remains so after Aveni and Hartung's new investigations published in 1986. There are still many unanswered questions. How was the Teotihuacan direction translated to such distant places to the Tropic of Cancer and to Guatemala? Why were there so many different alignments? What does the differing angles signify and why is it almost always a clockwise deviation? Are all the bearings connected with the calendar? How was the planned axis fixed at the place where a building was erected or a settlement established? And this require observations of the sun every time on a particular day, or were there procedures and implements to measure the angle? Today I shall try to answer the last question while touching upon others.

The flat or only slightly hilly region of North Yucatan makes it possible to observe sunrise and sunset on the horizon if there is no cloud. It is possible to fix the sunrise by using a pair of crossed sticks, for example, or with poles. The appropriate day is determined by the number of days to the next passage of the sun across the zenith. Observation of this is easy in the Tropics (Aveni and Hartung 1981; Tichy 1980: 132). Flat ground permits then very precise alignments with a margin of error of at least 1°. Even where the ground is not so flat, where mountains or a range of mountains obscure the horizon, as in the basins of Central Mexico, the various bearings are easy to distinguish. This raising of the horizon apparently did not have much effect on the patterns of buildings and settlements. Direct observation of the sun was not always undertaken to determine the required pattern.

Great importance was attached to astronomical observations for some especially significant buildings in ceremonial centers. Some large areas with their settlements and fields in the basins of Puebla-Tlaxcala (Tichy 1974), Mexico (Tichy 1976a, b) and Oaxaca (Storck 1980) also have precise bearings. Paths and field boundaries go in the same direction for very long distances until an area with a different bearing is reached. These are probably the former territories of tribes or rulers which are explicitly distinguished in this way (see fig. 1).

It is hardly likely that the sun was observed repeatedly on the relevant days in order to keep to the required direction. We can, however, assume that central raised points were used as observatories, from which exact observations were made so as to translate these alignments on to the surrounding land.

In large areas, such as the inner basin of Puebla, where Cholula determines the system of orientation, there were considerable deviations. This was not the case in smaller areas. For example, the Tepeaca-Tecamachalco plaine east of Puebla like Cholula is determined by the sunrise of the winter solstice (25–26°), but much more precisely (see fig. 2).

If a certain axial direction had to be fixed, without it being possible to observe sunrise or sunset on one of the four relevant days in the local horizon with no or little elevation, the angle must have been measured somehow. This requires a fixed point at the horizon, an agreed unit of measurement and some kind of device. We can assume that it was necessary and common to measure angles in Mesoamerica. But how was it done?

2. Ways to measure angles.

How could it have been done by the early nomadic people in the first stages of civilization? No suitable implement has been discovered by archaeologists. The precursor of an implement could have been a measure taken from the human body, like the units yard and foot. At first it was probably possible to indicate distance on the horizon using the extended arm with spread fingers. With the width of five spread hands I can cover a right angle, about 18° per hand.

Using a fixed orientation point, the top of a mountain as a topographical fixed point, or where the sun rises or sets, a solstice point, a direction can be indicated more or less accurately and communicated to others in a group of hunters or gatherers.

Using the individual spread fingers I can divide the 18° up into 5°, 9° or 13°. In this way with my outstreched arm I measure on the horizon or anywhere else an angle of a quarter of the open palm or a twentieth of the right angle – 4.5° or 5 decimal degrees (see fig. 9).

Since my first observations of oriented settlements in Central Mexico, I have assumed that this is the mesoamerican unit of angles. But I have as yet presented too little empirical evidence to support this hypothesis. I hope to do this today in order to refute the criticism made of my hypothesis.

Outside Mesoamerica the right angle has also been represented in twentieth parts and in relation to the human body – at a very early stage of human civilization. The Zürich engineer Amandus Weiss found one of these representations above the picture of a bison (see fig. 3) in the Lascaux cave. He believes that the nomadic hunters had a kind of compass 17,000 years ago which they could represent on the earth as soon as the north point had been fixed. Just above the bison a 25 cm wide arc of a circle goes across the whole length of the picture. It is made up of 12 to 15 fine short scratches one next to the other. On the picture it can be seen extending over 4 metres. From this arc there are 3.5 metre long thin lines equidistant from each other and running towards the centre of the arc. 3.5 metres is the equivalent of 5 steps each 70 cm long. The radial lines are 28 cm, i.e. one foot, apart when they reach the arc. In this quarter circle there are 20 of these radial lines, that is to say the right angle was divided into 20 parts and every footlength corresponded to exactly five decimal degrees or gradians (Weiss 1984: 316).

If a direction-determining implement of this kind was known to the very earliest hunters of Europe and beyond Europe in Eurasia, it is reasonable to assume that it was one of Man's first technical achievements in the northern hemisphere and that it was taken from Eurasia to America. So far no right angle divided into 20 parts or a circle divided into 80 parts has been found in North America although there are some circles with radials.

There are the "medicine wheels", such as the one in Big Horn, Wyoming which is 27 metres across and has 28 spokes (Eddy 1974). "Pecked circles" formed as rows of dots on stone have been discovered (Aveni, Hartung and Buckingham 1978; Aveni and Hartung 1980; Aveni, Hartung and Kelley 1982; see also fig. 4a and 4b). They could be regarded as symbols of a compass which angles were measured at this point. We are still looking for evidence for the use of the ancient angular measurement unit 4.5°, which is in accordance with the vigesimal system of Ancient America.

The angular measurement unit which I postulate resulted from the statistical treatment of directional values in the basins of Central Mexico. Beside the values of the relatively few and well surveyed pyramids I used the axial directions of the churches, the colonial successors of the temples in the highlands, which were settled in early times. These are at the same time the axial directions of the settlements, the villages with their fields. The histogram of the axes of 277 churches in the Mexico basin revealed that the axial directions are concentrated around astronomical east to south with the clockwise deviation 1–2° 7° 11–12° 16° 20° and 25° (see Tichy 1976a, b; see too fig. 5). The connection with the movement of the sun can be seen in the solsticial direction (25°), the bearings of sunset on the days of the passage of the sun through the zenith on 19° N.L. (20°) and the so-called midyear days near the equinoxes (1°). The 1°, 7° and 16° directions are explained by the first days of the periods of 20 days in a hypothetical fixed sun calendar posited according to the description by Sahagun for 1564–1567 in Tlatelolco (Tichy 1980). For several rare alignments in the highlands just as for 11° no explanation has as yet been found. It seems clear that the basis is not simply the calendar but also a geometric system that consists of arithmetical angular series with 4.5° intervals. The sequence of angles discovered in Central Mexico can have either the solsticial points in the east (90 + 25°) or in the west (270 + 25°) as fixed points on the horizon, because these are in every instance clockwise deviations. This fact reveals the great importance of these points and directions as cardinal points and cardinal directions of mesoamerican cosmology.

Using their own and data presented by other scholars relating to buildings, Aveni and Hartung (1986) have found the same angular values in some cases. In most cases, however, their values, collected in the Maya region, occur very rarely in the Highlands of Central Mexico. In addition to the deviations around 1° 7° 11° 16° and 25° already mentioned which occured less frequently, they found values of 5° 9° 14° 18° and 22° in the Maya area (see fig. 6). In common with the Mexico series, this Maya orientation series also has angular values deviating by 4° to 5°. This fact also lends support to the hypothesis that there was a mesoamerican angular measurement unit of 4.5°. Only the directions 1° 7° 9° 14° 16° are connected with the hypothetical solar calendar with 20 day periods – the Orientation Calendar according to Aveni and Hartung – but there is also an explanation covering both series at the same time – namely a hypothetical Solar Calendar with 13 day periods which is presented in the table (Tichy 1988b). I cannot here go into more detail about it but I assume that it is valid for, and can be applied to, the whole of Mesoamerica (see table and fig. 7).

An instrument is needed to measure angles and arcs in the sky or on the horizon more accurately. It should make it possible to measure differences between the positions of stars or between the points where the sun rises more accurately than with the outstreched arm with spread fingers. Perhaps this device looked like the strange object held by a figure in the Codex Selden (see fig. 8). Aveni (1980, fig. 7) calls it "a prolonged implement with stars perched on it". My reading of this picture is that it represents a hand holding out a tendril with 3 or 4 prongs or a bundle of twigs. The tips of the twigs are equidistant from each other. The stars suggest that the device was used for the observation of stars.

How far apart must the tips of the twigs be with the outstreched arm holding the device vertically to measure the assumed 4.5° angle? In my case with the distance between eye and instrument being about 55 cm the distance is = tan 4.5° × 55 cm = 4.3 cm (see fig. 9). With four tips the device would be 12.9 cm across at the top and would permit the observer to see 3/20 of the right angle or 13.5°. This is exactly how much that can be seen with the four spread fingers of the hand. This instrument, if it is one, seems to have calibrated or standardized the hand as a unit of measurement for observing angles.

With this device and using a flat piece of sufficently dark obsidian as protection for the eyes against the sun, it will have been possible to make other observations of the sun. The fact that below 19° N.L. at noon on June 21st, the summer solstice, the sun reaches its highest point – around 4.5° from the zenith towards the North – must certainly have been known and must have had considerable significance for such important ceremonial centres as Cholula, Cacaxtla and Xochicalco, which are close to this parallel of latitude.

An advanced civilization with such outstanding achievements in astronomy will not only have had suitable observatories such as the Caracol of Chichen Itzá and the building on Monte Alban (Aveni 1980: 249–286) or solar and calendar observatories such as the group E of Uaxactun or the group C and D of Xochicalco to fix the relationships of angles on the horizon (see Tichy 1985: 102). To plan and put up a building of any significance, to plan towns and settlements, including these in the open country, everything had to be made part of a system of orientation connected with an orientation calendar. Having ascertained that there was very probably a mesoamerican angular unit of 1/20th of a right angle, a suitable instrument can be reconstructed owing the illustrations in codices (Tichy 1988a). Perhaps one will be found one day although they were very probably made of wood which would rot. Even if this link in the chain of proof is missing, we have made considerable progress towards explaining the many alignments in the architecture of Mesoamerica and understanding the practical realisation of such plans.

In future when giving angles in connection with the cultures of Mesoamerica, we should use the new decimal graduation with the 400 decimal degrees (gradians) now employed in surveying in preference to the old 360° degree system. Measurement of the frontispeace of the Mendoza Codex which represents the legend of the foundation of Tenochtitlan produced a surprising result (see fig. 10). The diagonals cross at angles of 90^g and 110^g gradiens, that is 18 and 22 mesoamerican units. The diagonally opposite pairs of angles of the rectangle are 45^g and 55^g respectively, 9 and 11 of these units. This confirms me in my assumptions that the unit postulated was used. There is a clear connection with the proportions of the sides of the rectangle – 6 : 7. These proportions are found very frequently in Teotihuacan (the inner courtyard of the Ciudadela including the Quetzalcoatl Pyramid, the ground plan of the palaces on both sides of this pyramid, the courtyard in the buildings beyond Miccaotli (the Road of the Dead) and in several courtyard complexes in Tetitla). The base of the Quetzalcoatl Pyramid in Xochicalco also has the same proportion 6 : 7. Other rectangular ground plans, steles and hieroglyph stone slabs have diagonal angles whose rounded values can be expressed in gradians. When the sides are in proportion 1 : 2 it is 30^g, 1 : 3 20^g, 5 : 7 40^g. I suggest that it would be worthwhile to follow up these relationships.

References

Aveni, A.F.
1980 *Skywatchers of Ancient Mexico.* University of Texas Press, Austin and London.

Aveni, A.F., S.L. Gibbs and H. Hartung
1975 The Caracol Tower at Chichen Itza: An Ancient Astronomical Observatory? *Science* 188: 977-985.

Aveni, A.F. and H. Hartung
1976 Investigación preliminar de las orientaciones astronómicas de Copán. *Yaxkin* 1: 8-13. Tegucigalpa. Instituto Hondureño de Antropología e Historia.

Aveni, A.F. and H. Hartung
1981 The Observation of the Sun at the Time of Passage through the Zenith in Mesoamerica. *Archaeoastronomy* Nr. 3, Supplement to Journal for the History of Astronomy 12: 51-70.

Aveni, A.F. and H. Hartung
1986 Maya City planning and the Calendar. *Transactions of the American Philosophical Society* 76. Part 7: 1-87.

Aveni, A.F., H. Hartung and B. Buckingham
1978 The Pecked Cross Symbol in Ancient Mesoamerica. *Science* 202: 267-279.

Aveni, A.F., H. Hartung and J.Ch. Kelley
1982 Alta Vista (Chalchihuites), Astronomical Implications of a Mesoamerican Ceremonial Outpost at the Tropic of Cancer. *American Antiquity* 47: 316-335.

Eddy, J.A.
1974 Astronomical Alignment of the Big Horn Medicine Wheel. *Science* 184: 1035-1043.

Storck, K.-A.
1980 Die Orientierung von Orts- und Flurgrundrissen im Becken von Oaxaca/Mexiko. *Lateinamerika-Studien* 6: 139-163. München.

Tichy, F.
1974 Deutung von Orts- und Flurnetzen im Hochland von Mexiko als kultreligiöse Reliktformen altindianischer Besiedlung. *Erdkunde* 18: 194-207. Bonn.

Tichy, F.
1976a Ordnung und Zuordnung von Raum und Zeit im Weltbild Altamerikas. Mythos oder Wirklichkeit? *Ibero-Amerikanisches Archiv* N.F. 2: 113-154. Berlin.

Tichy, F.
1976b Orientación de las pirámides e iglesias en el Altiplano Mexicano. *Suplemento Comunicaciones* IV: Puebla. Fundación Alemana para la Investigación Científica.

Tichy, F.
1980 Der Festkalender Sahagun's. Ein echter Sonnenkalender? In *Wirtschaft und gesellschaftliches Bewußtsein in Mexico seit der Kolonialzeit,* edited by H.-A. Steger und J. Schneider, pp. 115-137. Lateinamerika-Studien 6, Wilhelm Fink Verlag, München.

Tichy, F.
1976/81 Order and Relationships of Space and Time in Mesoamerica: Myth or Reality? In *Dumberton Oaks Conference on Mesoamerican Sites and World Views,* edited by E.P. Benson, pp. 217-245. Washington D.C. 1976, Copyright 1981.

Tichy, F.
1982 The Axial Direction of Mesoamerican Ceremonial Centers on 17° North of West and their Associations to Calendar and Cosmovision. In *Space and Time in the Cosmovision of Mesoamerica,* edited by F. Tichy, pp. 63-83. Lateinamerika-Studien 10, Wilhelm Fink Verlag, München.

Tichy, F.
1985 Sonnenbeobachtungen und Agrarkalender in Mesoamerika. In *Gedenkschrift für Gerdt Kutscher* Teil 2, edited by A. Mönnich, B. Riese und G. Vollmer, pp. 99-112. Indiana 10. Berlin.

Tichy, F.
1988a Der Fuß zur Längenmessung in Mesoamerika, die Hand zur Winkelangabe? In *Festschrift für Thomas Barthel zum 65. Geburtstag,* edited by V. Harms. Tübingen.

Tichy, F.
1988b Orientierungskalender in Mesoamerika. Hypothesen über Aufbau, Anwendung und Verbreitung. (Ms.).

Weiss, A.
1984 Orientierung der Wanderjäger im Paläolithikum. *Naturwissenschaftliche Rundschau* 37: 312-319.

Hypothetical Solar Orientation Calendar with 13-day Weeks beginning with Winter Solstice and the sequences of angles with the difference of 4,5°.

1st day of 13-day week	astronomical events agrarian cycle		clockwise deviation from W or E sunset in summer half-year sunrise in winter half-year	Lat. 15°N	19°N	20,5°N	
22/12	Winter Solstice	WINTER H.-Y.	**25**	24°19'	24°53'	25°08'	
4/1				23°35'	24°07'	24°22'	
17/1			**22**	21°32'	22°01'	22°14'	
30/1			**18**	18°20'	18°44'	18°55'	
12/2	Beginning Agrarian Cycle		**14**	14°12'	14°32'	14°40'	◁
25/2			**9**	9°27'	9°39'	9°45'	◁
10/3			**4**	4°17'	4°23'	4°25'	◁
23/3	Mid-year Day (Vern. Equinox)		**1**	1°01'	1°03'	1°03'	
5/4		SUMMER-HALF-YEAR	**6**	6°15'	6°23'	6°27'	◁
18/4			**11**	11°11'	11°25'	11°32'	◁
1/5	1. Solar Zenith Passage Lat. 15°N		**16**	15°36'	15°56'	16°06'	◁
14/5			**20**	19°17'	19°43'	19°55'	◁
27/5	1. Solar Zenith Passage Lat. 21°16'			22°04'	22°34'	22°48'	}
9/6				23°47'	24°20'	24°34'	}◁
22/6	Summer Solstice		**25**	24°19'	24°53'	25°08'	}
5/7				23°39'	24°12'	24°26'	}
18/7	2. Solar Zenith Passage Lat. 21°16'N			21°49'	22°19'	22°32'	}◁
31/7			**20**	18°58'	19°23'	19°35'	}
13/8	2. Solar Zenith Passage Lat. 15°N		**16**	15°14'	15°34'	15°43'	◁
26/8			**11**	10°49'	11°04'	11°10'	◁
8/9			**6**	5°56'	6°04'	6°07'	◁
21/9	Mid-year Day (Aut. Equinox)			0°46'	0°47'	0°48'	◁
4/10		WINTER H.-Y.	**4**	4°28'	4°33'	4°36'	◁
17/10			**9**	9°33'	9°45'	9°51'	◁
30/10	Conclusion Agrarian Cycle		**14**	14°15'	14°34'	14°42'	◁
12/11			**18**	18°20'	18°43'	18°54'	◁
25/11			**22**	21°30'	21°59'	22°12'	◁
8/12				23°33'	24°06'	24°20'	
	21/12 one day rest						

Winter (Maya) Orientation Serie: 4°- 9°- 14°- 18°- 22°- 25°

Summer (Central Mexico) Orientation Serie: 1°- 6°- 11°- 16°- 20°- 25°

Triangles: Difference 4°- 5°

Calculation based on a horizon altitude of 0,5° and an atmospheric refraction of 0,5°. The position of sunrise or sunset will appear on the astronomical horizon A_o.

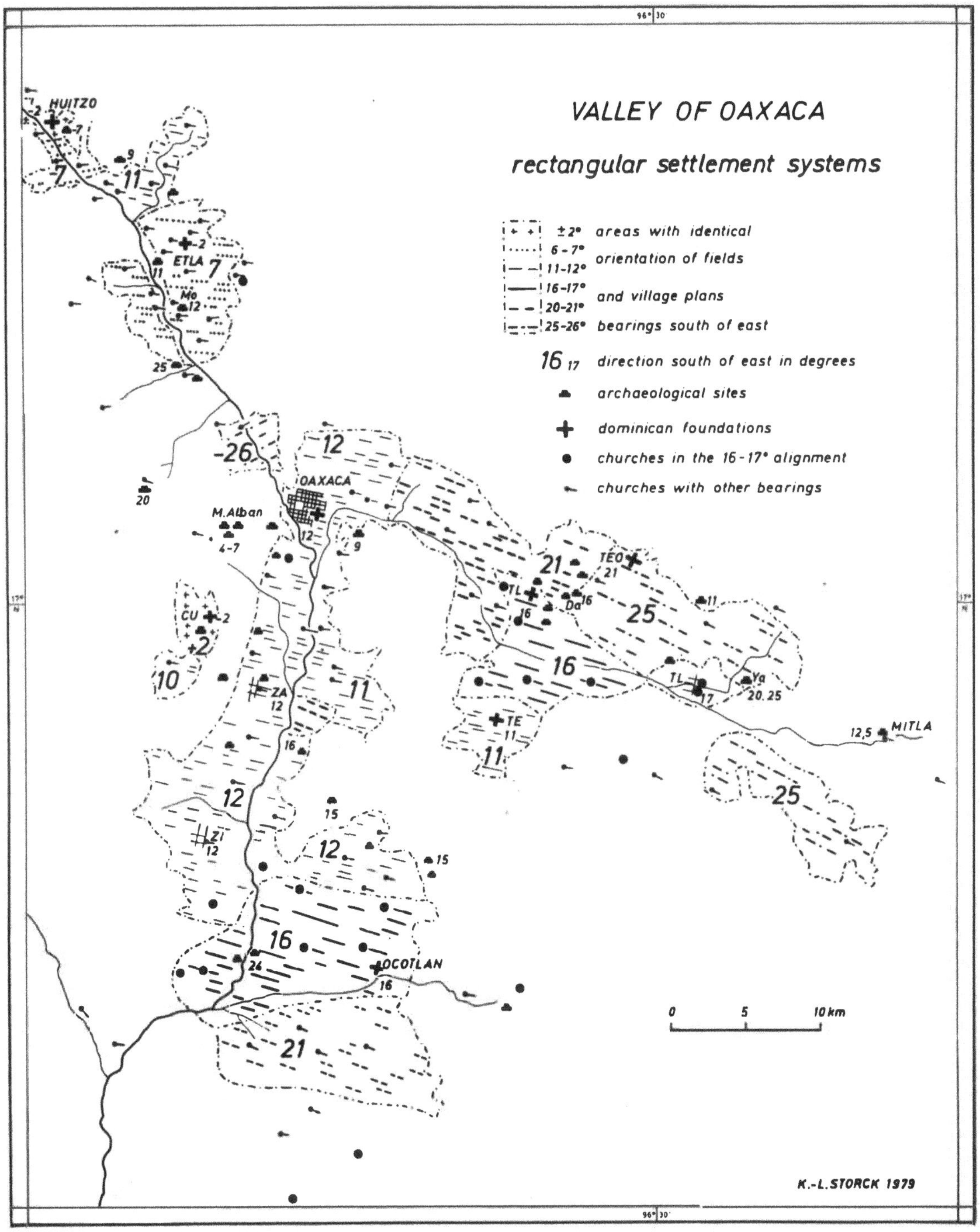

Fig. 1: The rectangular settlement system in the Valley of Oaxaca (after Storck 1980).- Numbers indicate the deviation in degrees; points and straight lines: settlements and fields in the 16–17° alignment; crosses: dominican foundations and their bearings. (Tichy 1982, Fig. 5).

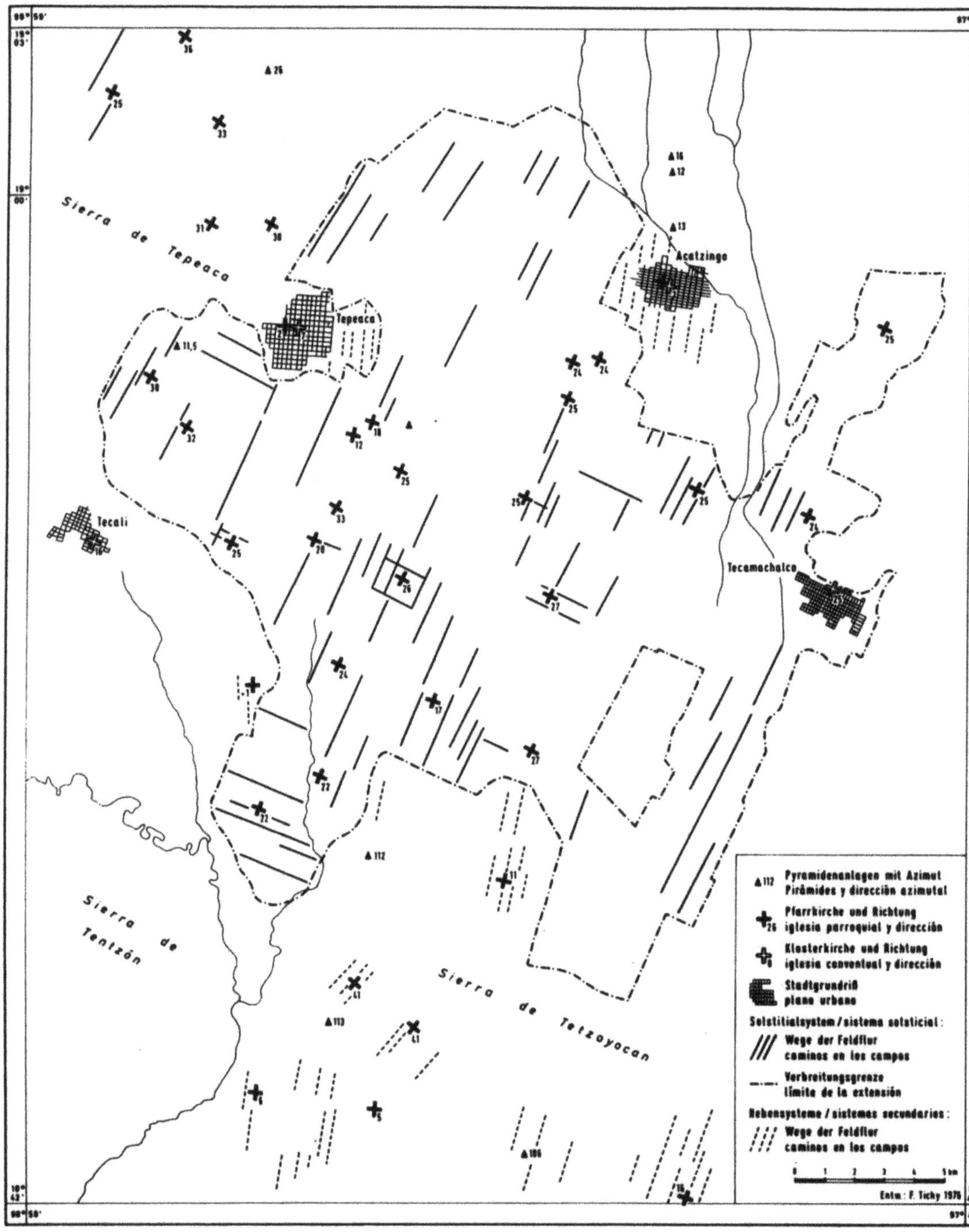

Fig. 2: Settlement pattern in the plain of Tepeaca-Tecamachalco/Puebla with its alignments towards sunrise on the day of the winter solstice E 25° S (after Tichy 1988a).

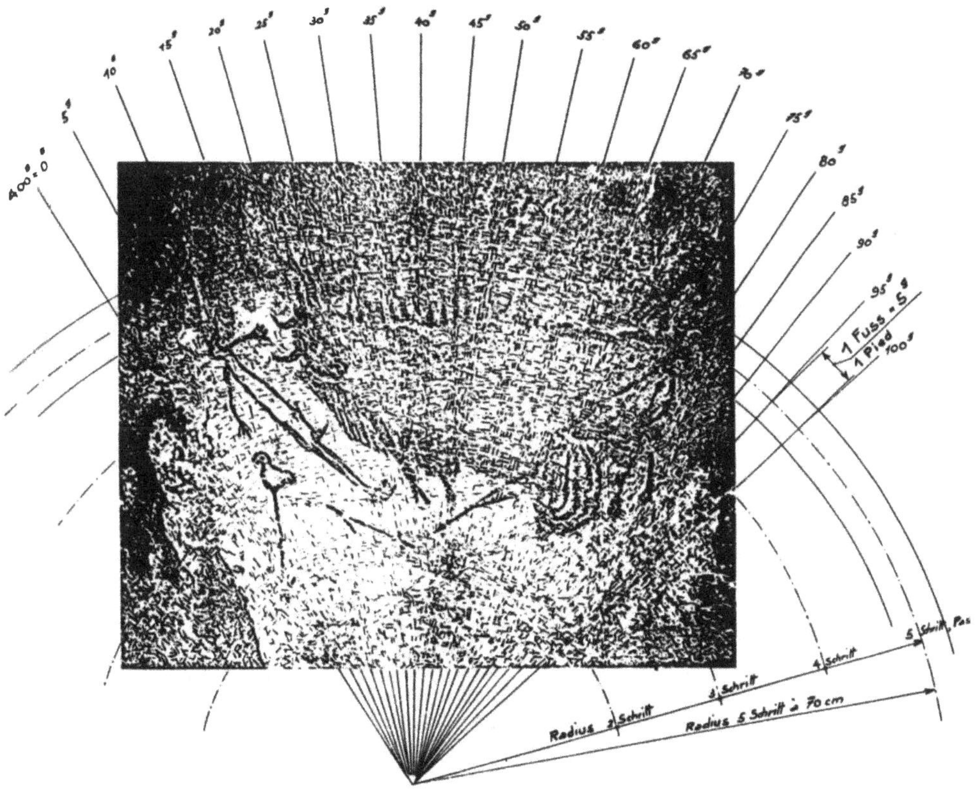

Fig. 3: Compass of the eolithic nomadic hunters in the Lascaux cave, France, after Weiss 1984, ill. 4. An arc has been drawn above the picture of a bison using fine lines. The arc has a radius of 5 steps each of 70 cm. The right angle is divided into 20 parts each of 5 gradians, which on the arc are each one foot long (28 cm).

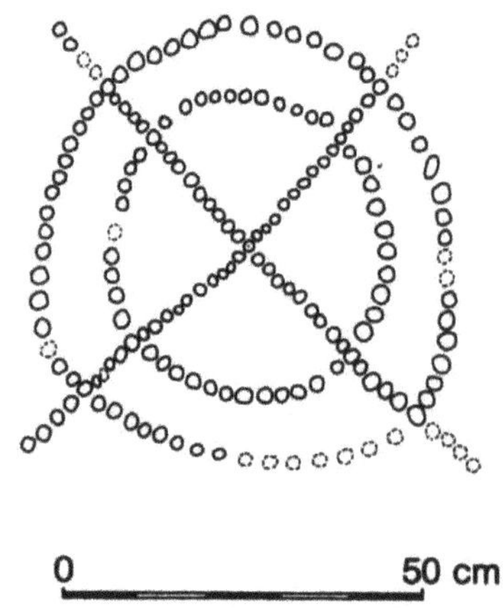

Fig. 4a: Circle pecked on vertical outcrop near Tepeapulco 33 km northeast of Teotihuacan facing Cerro Gordo. Diameter 58 cm. The inner quadrants contain 9 – 10 – 11 – 12 holes, the outer quadrants 16 – 14 – 15 – 15 holes. After Aveni, Hartung and Buckingham 1978: Fig. 1h.

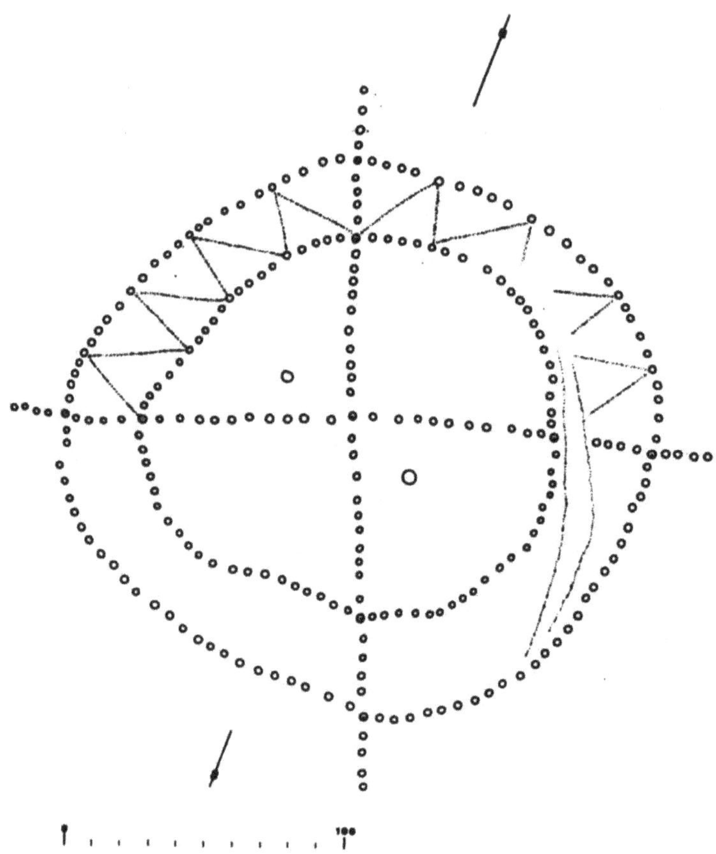

Fig. 4b: Pecked circle on plateau of Cerro El Chapin, 23° 24′ N near Alta Vista. Diameter 200 cm. The inner quadrants contain 20, the outer quadrants 25 holes, without the holes of the axes. After Aveni, Hartung and Kelley 1982: Fig. 4b.

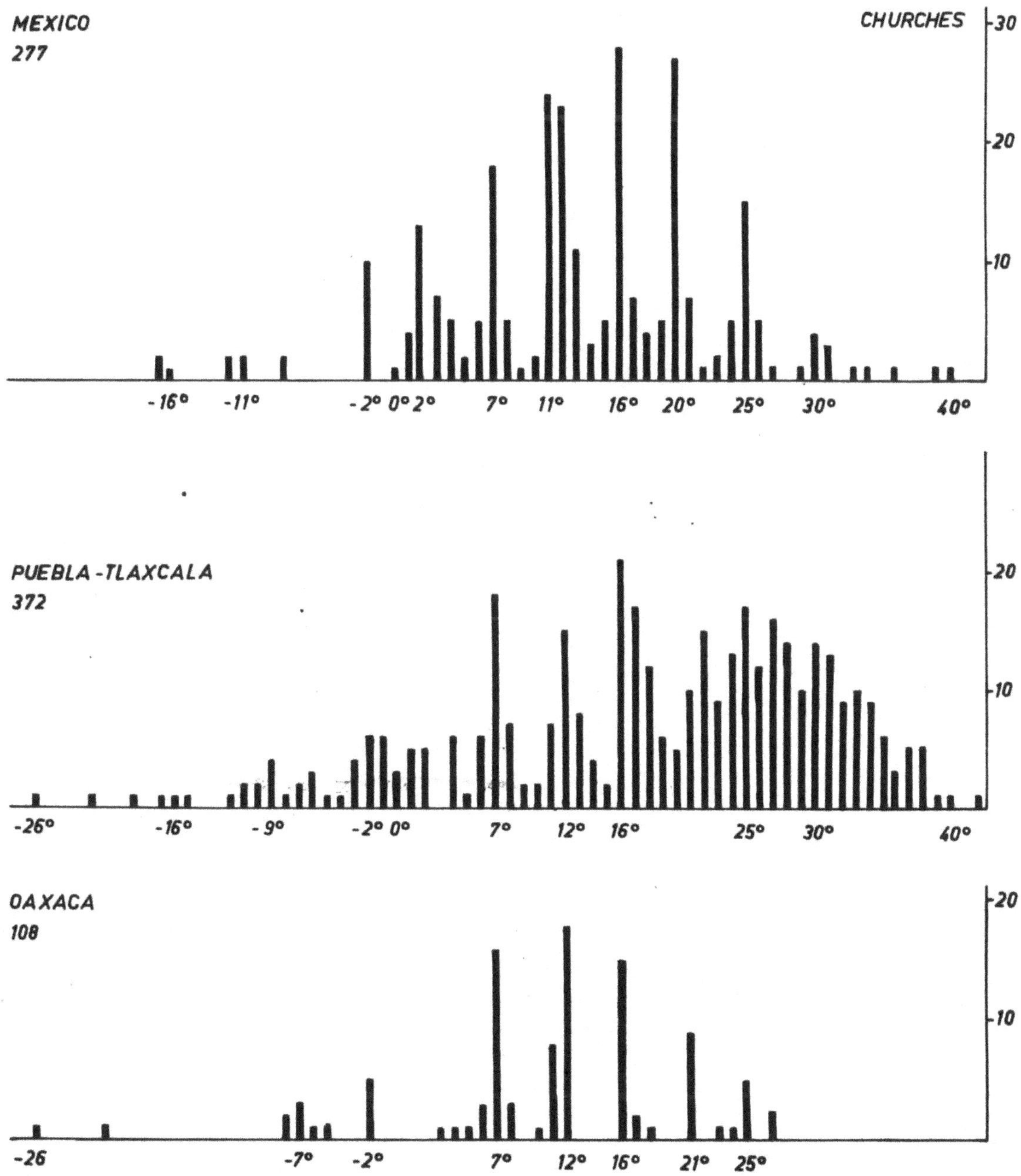

Fig. 5: Histograms showing the distribution about astronomical east of the axes of 277 churches in the Mexican basin, of 372 churches in the Puebla-Tlaxcala-Basin and of 108 churches in the Oaxaca-Valley. - From Tichy 1982: 65.

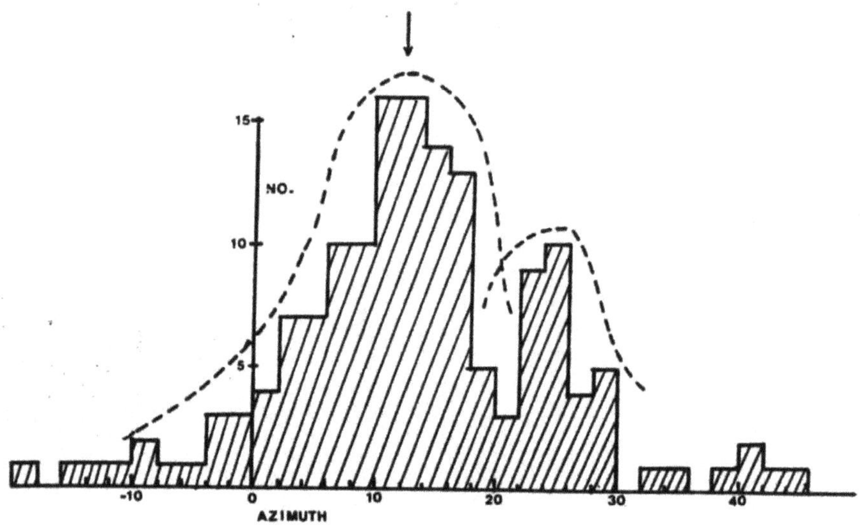

Fig. 6: Frequency diagram of orientation lines of important buildings in the Maya area. Transit measurements by Aveni and Hartung in 1971, 1973, 1976, 1980, 1984 and 1986. - From Aveni and Hartung 1986 Fig. 2a. - The arrow marks the median direction.

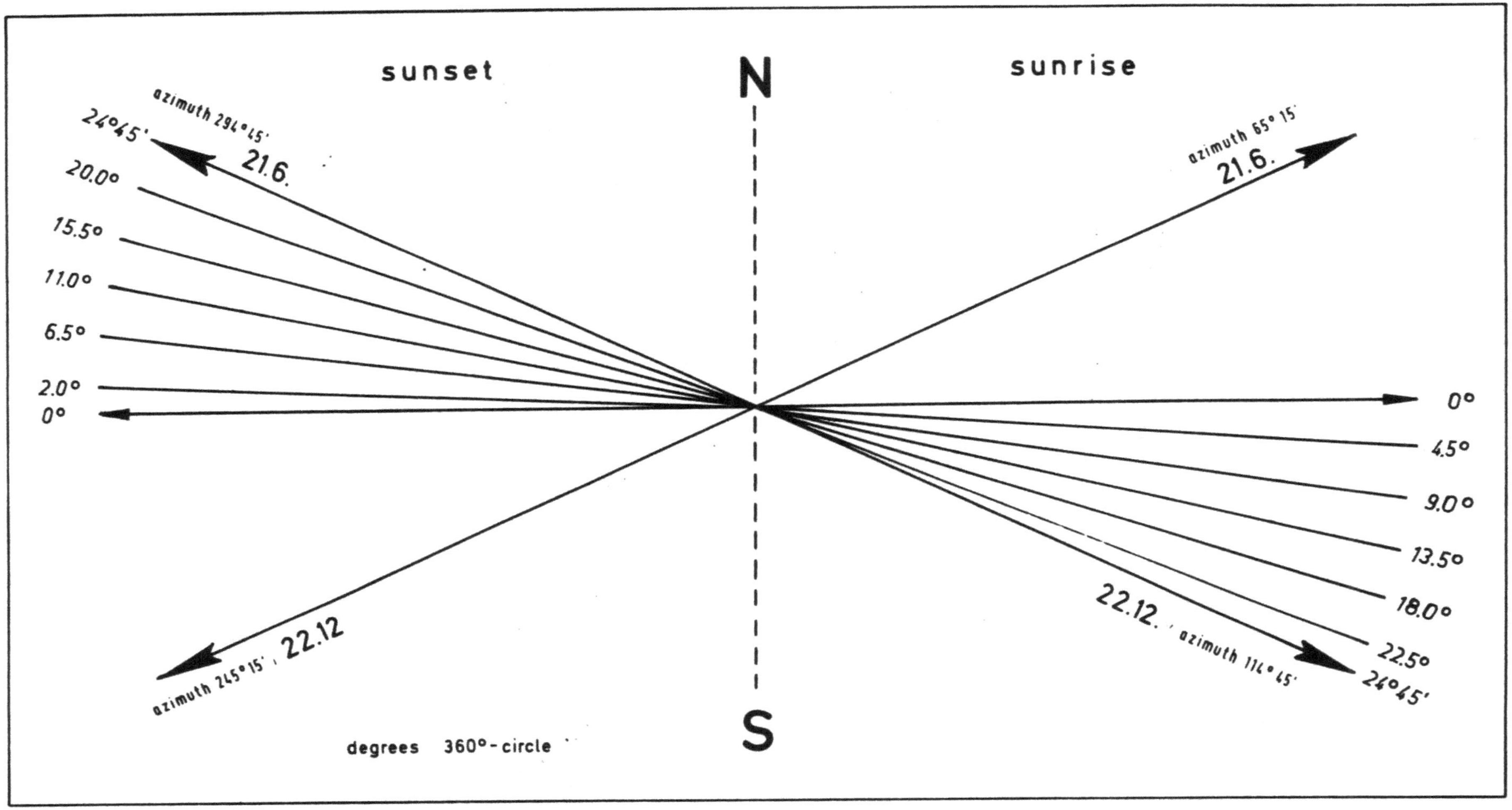

Fig. 7: Diagram of the Mesoamerican Orientation System. The alignments are in accordance with the hypothetical solar orientation calendar with 13-day weeks. The clockwise deviations from East represent the "Winter Sunrise Maya Type" and the deviations from West the "Summer Sunset Centralmexico Type".

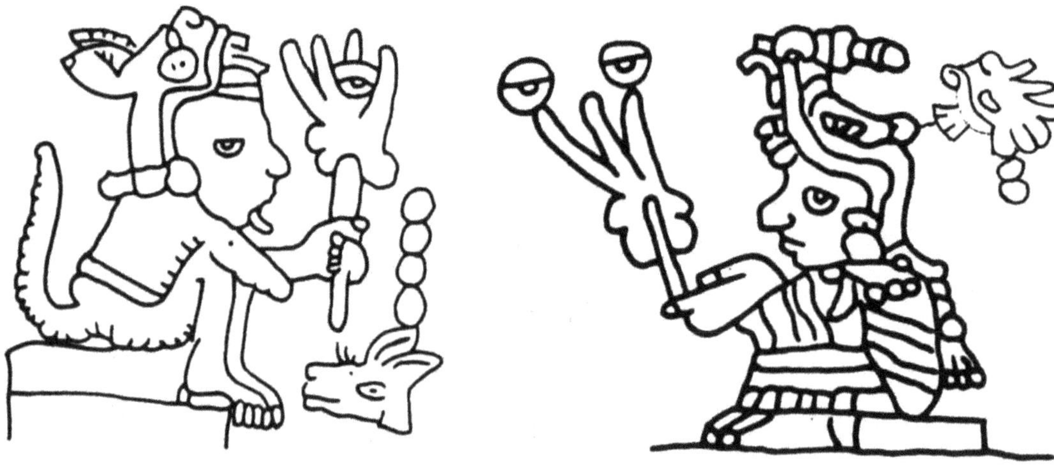

Fig. 8: Person with an implement which can be interpreted as being a device used for astronomical observation.

Left: Man 4 Deer: Codex Selden p. 14–IV, holding implement which 4 prongs.- "Jaguar-torch with eye" after A. Caso, Reyes y reinos de la Mixteca. Vol. 2, México 1979, p. 164.

Right: Woman 2 Water: Codex Bodley p. 17–IV, holding implement with three prongs.- "Fire snake torch stars" after A. Caso, Interpretación del Códice Bodley 2858. México 1960, p. 17. After Tichy 1988a, Fig. 5.

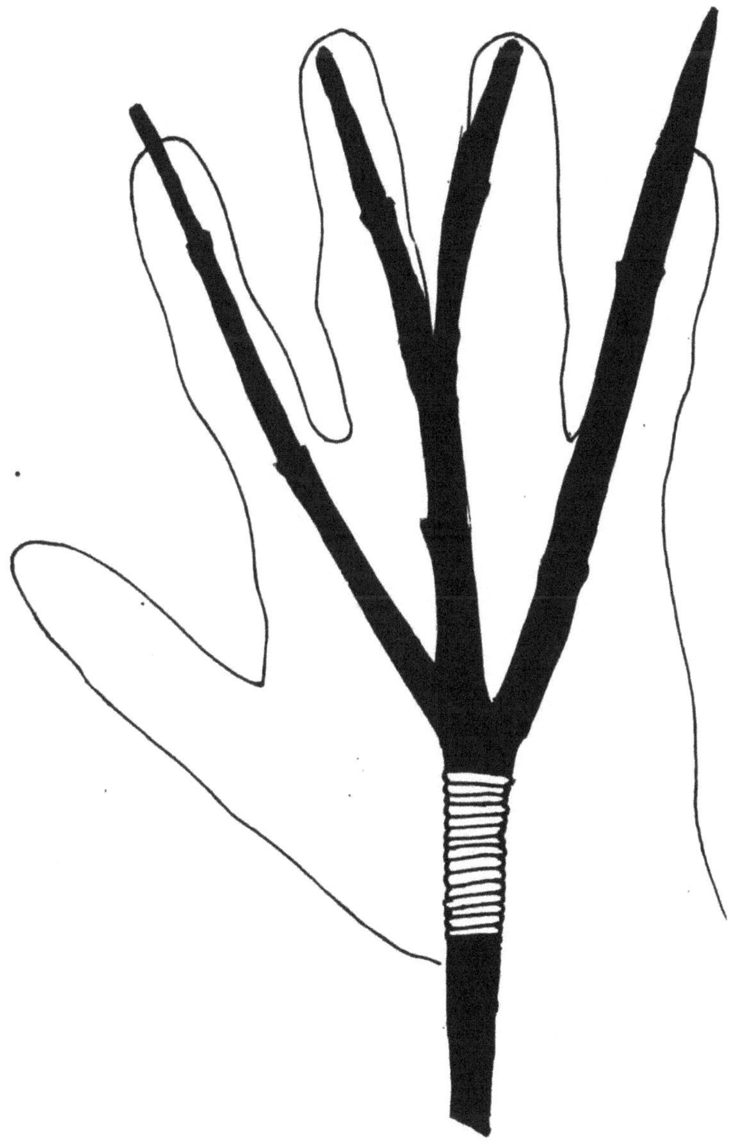

Fig. 9: Reconstruction of an astronomical observation device based on an illustration in Codex Selden. It is the size of a hand with spread fingers and made of a bundle of twigs. - When the hand is outstreched the angle between two tips which are 4,3 cm apart is 4.5° or 1/20th of a right angle (sketch by Tichy); see Tichy 1988a.

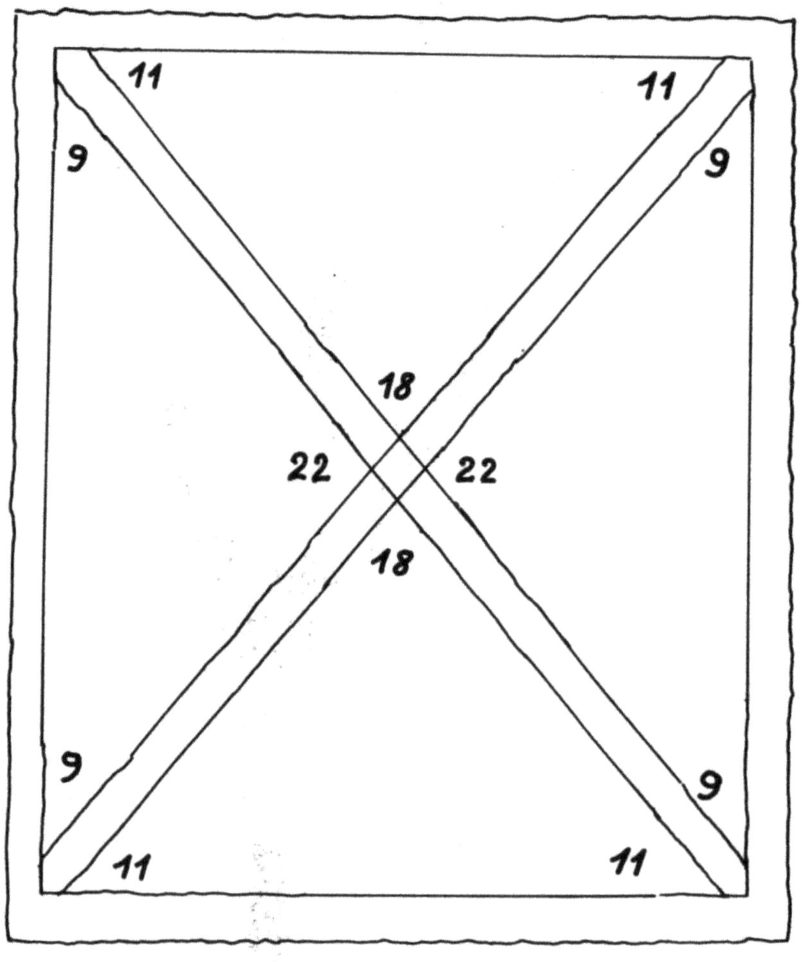

Fig. 10: Frontispiece of Codex Mendoza representing the foundation of Tenochtitlan with the crossing water courses and four parts of the town in the rectangle of Lake Texcoco. - Proportions between the sides 6:7. Angles given as multiples of 4.5° or 5^g = 1/20th of right angle, the assumed mesoamerican unit of measurement.

Astronomía Primitiva entre los Taínos y los Caribes de las Antillas

Sebastián Robiou-Lamarche

Al momento del Descubrimiento existían dos grupos culturales principales que ocupaban las islas antillanas, resultado de una serie de migraciones arahuacas y caribes provenientes del área de la cuenca del Orinoco así como de un desarrollo propiamente regional.

Las Antillas Mayores, constituídas hoy por Cuba, Jamaica, La Española (Haití- República Dominicana) y Puerto Rico, estaban pobladas por los taínos, cultura antillana de origen arahuaco, mientras las Antillas Menores constituídas por islas más pequeñas como Antigua, Guadalupe, Dominica, Martinica, Santa Lucía, Barbados y Granada, estaban ocupadas por las últimas migraciones caribes provenientes también de las costas de Suramérica. (Figura 1). Las diferencias culturales entre ambos núcleos poblacionales ha sido motivo de diversas polémicas a través de los años (Sued Badillo, 1978; Alegría, 1981). No obstante, el máximo desarrollo cultural antillano hacia la época del Descubrimiento puede establecerse, sin dudas, en la sociedad caciquil taína que ocupaba la Española y Puerto Rico (Moscoso, 1986).

Debido al impacto de la conquista y la colonización española, la cultura taína de las Antillas Mayores y las Bahamas desapareció prácticamente en menos de cien años. La visión que hoy tenemos de la sociedad taína se debe a las limitadas descripciones que hicieran Colón (1961, 1962), Martir de Anglería (1964), Pané (1974), Las Casas (1909, 1951) y Oviedo (1959). Por su parte, la cultura caribe isleña logró en cierto grado perdurar hasta mediados del siglo XVII cuando se estableció la colonización francesa en las Antillas Menores. Cronistas franceses como Bouton, Coppier, Rochefort, Breton, La Borde y otros (Cárdenas Ruíz, 1981) nos permiten conocer parte de la mitología, leyendas y creencias de los caribes isleños.

La investigación de la astronomía primitiva en las Antillas, por lo tanto, carece de la documentación etnohistórica que quisieramos tener. Sin embargo, la evidencia existente nos permite pensar que, al igual que en el resto de América, la astronomía estaba íntimamente integrada a la mitología, las manifestaciones artísticas, las construcciones, las prácticas chamánicas, la estructura social, los ritos, la confección de calendarios, en fin, en toda la cosmovisión indígena antillana.

El propio Almirante Cristobal Colón fue quizás el primero en constatar la importancia de la astronomía entre los taínos. (Figura 2). Las Casas (1951) nos narra cuando Colón se encontraba encallado con sus carabelas destruídas en Jamaica, durante el transcurso del cuarto viaje. Los indios se negaban suplirle los abastecimientos necesarios para él y sus hombres. Entonces, conociendo que dentro de unos días ocurriría un eclipse lunar, Colón llamó a los jefes o caciques señalándoles que su acción sería castigada por Dios con cierta señal en el cielo. La noche del 29 de febrero de 1504 los atemorizados indios no dudaron de la conección divina de Colón supliéndole de ahí en adelante con toda la comida necesaria hasta su posterior rescate. Otros cronistas españoles nos refieren breves citas que señalan la importancia simbólica de la Luna para los taínos. En La Española los taínos sembraban la yuca (Manihot esculenta Crantz) o mandioca "...siempre después que la luna ha hecho o se muestra nueva, o les presto que ser puede en los días que crece hasta el lleno della, pero nunca en menguante", según el cronista Oviedo (1959: 227). También la Luna servía para un aparente calendario lunar pues según Mártir de Anglería (1964, capítulo III) escribe: "Hizose pacto que los habitantes de los montes enviarían a la ciudad cada tres meses, que ellos por La Luna llaman lunas, cierta medida de oro que les fue señalada..."

Sin embargo, será el análisis de la mitología taína y el estudio arqueoastronómico de las llamadas plazas ceremoniales taínas los que nos permitirán postular la existencia de una astronomía primitiva en las Antillas Mayores.

La fuente original de nuestros conocimientos sobre la mitología taína es la obra Relación Acerca de las Antigüedades de los Indios de fray Ramón Pané, escrita por mandato de Colón a partir del segundo viaje y cuyo manuscrito parece que terminó de escribirse hacia 1498. Pané refiere que los taínos creían que el Sol y la Luna salieron de una cueva. Por ello, en una cueva sagrada guardaban dos ídolos o cemíes hechos de piedra, pequeños, con las manos atadas. Más que el Sol o la Luna, Máhoru parece que representaba el principio solar relacionado al día, la sequía y el cacique o autoridad máxima, mientras Boínayel parece que representaba el principio lunar asociado a la noche, las aguas y el behíque o chamán taíno. El equilibrio y el control entre estos poderes gemelos era lo más importante para evitar el desbalance en la Naturaleza, fuera por extrema sequía o grandes inundaciones.

De allí que ambos ídolos fueran de igual tamaño y tuviesen "las manos atadas", quizás desatando en un rito las manos del cemí necesario en determinadas épocas del año. (Figura 3).

Nuestro estudio de la mitología taína en los últimos años, utilizando el análisis estructural y el estudio comparativo con mitologías meso y suramericanas, nos ha permitido encontrar una serie de asociaciones y relaciones astronómica-meteorológicas muy interesantes (Robiou, 1983 a, 1983 b, 1984 a, 1985 a).

De acuerdo a la fragmentada mitología recopilada por Pané, la deidad Yaya, equivalente al "Sumo-Espíritu" sin origen (Arrom, 1974: 65), tuvo un hijo llamado Yayael a quien desterró y finalmente mató poniendo sus huesos en una calabaza que colgó del techo de la vivienda. De los huesos de Yayael surgirían luego los peces y el mar. En muchos mitos suramericanos los peces y el mar están contenidos dentro de un gran árbol, especie de árbol de la vida o eje del mundo (*axis mundi*), del cual brotan por la intervención de un héroe mítico (Levi-Strauss, 1968). En otra serie de mitos suramericanos se ha verificado que la mutilación del héroe "resulta directa o indirectamente en la abundancia de peces en el agua y, en el cielo nocturno, la presencia de la constelación de Orión" (Lévi-Strauss, 1970). La vivienda taína, llamada bohío, era típicamente redonda con techo en forma de campaña sostenido por un gran poste central. Este poste central, de donde cuelga la calabaza mítica conteniendo los peces y el mar, tiene un gran paralelismo con el eje del mundo o el árbol de los alimentos a que hemos hecho referencia. A su vez, los poblados indígenas en las Antillas parece que también eran trazados siguiendo un patrón en el ordenamiento del espacio. Por lo menos en la región de Higüey, en La Española, el cronista Las Casas (1951) indica que los poblados eran construídos con una plaza central y cuatro especie de calles en forma de cruz. Es posible que esta arquitectura primitiva antillana tuviera relación con la astronomía, según ocurre en muchas tribus suramericanas (Magaña & Mason, eds., 1986).

El propio Colón (1961) en su *Diario* reporta haber visto en los bohíos, durante su primer y segundo viaje acaecidos en otoño, calabazas o cestillos con huesos colgadas del poste de la vivienda. La acción era obviamente un rito taíno que re-creaba, rememoraba, el mito de Yayael. Era, en otras palabras, un rito funerario propiciatorio de buena y abundante pesca. La observación de Colón coincide con la época del año en la cual los sábalos, róbalos y lisas, los peces más importantes en la dieta indígena, eran capturados en las desembocaduras de los ríos. Esta época de desove coincide con la aparición de Orión vespertino en el horizonte oriental hacia principios de diciembre, el final de la época de lluvias y el inicio de la época de sequía en las Antillas. Cabe pensar, pues, que similar a muchas tribus suramericanas, entre los taínos antillanos existiera una correlación entre la abundancia de peces en los ríos, la época de sequía y la aparición de Orión.

Otro mito taíno con implicaciones astronómicas-meteorológicas es el de los niños-hambrientos-llorones convertidos en ranas. Por Pané sabemos que el héroe Guahayona inicia su viaje mítico en canoa llevándose todas las mujeres consigo y dejando a los niños solos junto a un arroyo. Los niños, al faltarles sus madres, comienzan a llorar por el hambre diciendo "toa, toa", como quien pide una cosa con gran deseo y muy despacio se fueron transformando en ranas. En muchas partes de América los niños mitológicamente están relacionados con las estrellas o Pléyades. Según Krickeberg (1971), los "cuatrocientos-muchachos" del Popul Vuh son idénticos a las "cuatro-cientas-serpientes-de-nubes" de los mitos mexicanos, pues finalmente se convierten en estrellas o las Pléyades. Por otro lado, en un extenso grupo de mitos suramericanos, un grupo de niños-hambrientos-ruidosos-llorones ascienden lentamente al cielo convirtiéndose en estrellas o específicamente en las Pléyades. Lévi-Strauss (1968) se ocupó extensamente de este complejo mitológico en su obra Lo Crudo y lo Cocido. Las Pléyades son para algunas tribus suramericanas las estrellas que "hacen crecer la yuca o la mandioca" y son, para gran parte del Orinoco y las Guayanas, las estrellas que marcan el inicio del año o señalan la época de lluvias (Roth, 1915).

La estructura del referido mito taíno es idéntica a la de muchos mitos suramericanos, solo que un símbolo se ha transformado. La rana vendría a ser, como en casi toda América tropical, un símbolo de lluvia. Ahora bien, ¿existe una relación entre las Pléyades y las lluvias a nivel antillano? Las Pléyades vespertinas desaparecen en el horizonte occidental hacia el 26 de abril en la época del 1500 para la latitud de las Antillas. Esta época coincide con el inicio del ciclo de lluvias el cual logrará su máximo durante los meses de mayo-junio y termina hacia mediados de octubre. La desaparición de las Pléyades podría marcar la fecha de inicio de la siembra de la yuca para lograr que el retollo estuviera robusto antes de las fuertes lluvias de mayo y junio. La reaparición de las Pléyades vespertinas tiene lugar hacia finales de mayo extendiéndose este ciclo matutino hasta finales de octubre, es decir, durante toda la época de lluvia en las Antillas. Entre los caribes isleños, varios cronistas señalan a las Pléyades como las estrellas que marcaban el inicio del año (Cárdenas Ruíz, 1981). Aunque no existe esta evidencia para los taínos, es muy posible que estos también utilizaran las Pléyades para establecer un calendario agrícola-ceremonial. (Figura 4).

Por tanto, la transformación mitológica de las Pléyades en ranas entre los taínos vendría a decirnos que ambos símbolos tenían la misma identidad. Las ranas, originadas de los niños-llorones-hambrientos, son símbolo del agua augurada por las Pléyades, íntimamente relacionadas al ciclo agrícola. Cabe señalar que el símbolo de la rana está ampliamente extendido en todo el arte taíno. (Figura 5).

Si los dos mitos taínos analizados recuerdan el origen suramericano de esta cultura antillana, el tercer mito que veremos nos evoca parte de la mitología mesoamericana y el principio mitológico del axis mundi reportado en gran parte de América tropical (Aveni & Urton, eds., 1982). Según Pané nos narra en su Relación, Guahayona se llevó todas las mujeres en su canoa luego de salir de la cueva mítica origen de los taínos y dejar- como vimos- a los niños junto a un arroyo. Pero antes de esta acción ocurrió algo muy significativo: el héroe navegante taíno se deshizo de su cuñado Anacacuya, lo que le permitió iniciar su mítico viaje. Para ello, ya en la canoa, le enseña un hermoso cobo o caracol que ve en el agua. Aprovechando el descuído, Guahayona tomó por los pies a Anacacuya lanzándolo al mar. Pues bien, de acuerdo a Arrom (1974), Anacacuya significa "Espíritu-Central" o "Lucero-del-Centro". Sabemos que el centro alrededor del cual parecen girar todas las estrellas en el hemisferio norte corresponde a la estrella Polar, debido a que el eje de la Tierra apunta prácticamente hacia esa estrella, convirtiéndola en el Norte astronómico o único punto fijo en la bóveda celeste. Este hecho constituye el fundamento de toda la navegación primitiva. Por tanto, la acción de tomar a Anacacuya por los pies y lanzarlo al mar, convirtiéndolo en el "Lucero-del-Centro", es decir, en la Polar, es lo que permite al héroe nauta Guahayona iniciar su travesía en canoa.

Como sabemos, la Polar está también asociada a la Osa Mayor. La Osa Mayor, a su vez, en el marco del mar Caribe está directamente relacionada con el destructor fenómeno meteorológico llamado huracán, según demostró a principios de siglo Lehmann-Nitsche (1924), vocablo que claramente proviene de la deidad maya-quiché Hurakán, personaje de una sola pierna y según el Popul Vuh, "Centro-del-Cielo". Surge así un paralelismo simbólico entre la deidad maya-quiché y la taína: ambas están asociadas a una pierna o a los pies y ambas tienen una estrecha relación con el eje del mundo o la estrella Polar. Para los caribes isleños, vale señalar que la Osa Mayor también era una constelación relacionada con la canoa y la época de truenos y fuertes lluvias, pues según Taylor (1946) era llamada "lukúni-yábura" o "la-canoa-de-la garza".

Ahora bien ¿a qué se debe la identidad de la Osa Mayor

con el destructor huracán que ocurre en las Antillas y la península de Yucatán principalmente durante los meses de agosto y septiembre? Si observamos el ciclo anual de la Osa Mayor alrededor de la Polar notaremos que esta constelación parece sumergirse, entrar de "cabeza" al mar con lo que sería "la pierna" o "pies" hacia arriba, a mediados de abril. Para la latitud de las Antillas y Mesoamérica, la Osa Mayor desaparece bajo el horizonte por unos cuatro meses a partir de esa época y es hacia mediados de agosto y durante septiembre cuando emerge en el horizonte oriental al amanecer. (Figura 6). En otras palabras, la desaparición de la Osa Mayor bajo el horizonte, "dentro" del mar, coincide con la época de lluvias antillana y su reaparición en el horizonte claramente señala los meses de mayor ocurrencia de huracanes en las Antillas y el Golfo de México. Es, por lo tanto, de esta relación mitológica-astronómica que surge la identidad simbólica entre la Osa Mayor, la deidad maya-quiché Hurakán, el Anacacuya taíno y el "lukúniyábura" caribe isleño con el temible fenómeno meteorológico que hoy lleva el nombre de huracán. Y es posiblemente la observación del ciclo de la Osa Mayor alrededor de la Polar que el primitivo antillano representara este principio de fuerza cósmica en el símbolo de la espiral y otros símbolos sigmoideos, según estudió el cubano Fernando Ortíz (1947).

Pero la relación entre el mito taíno y el maya-quiché se estrecha aun más cuando recordamos que según el Popul Vuh la deidad Hurakán tenía tres manifestaciones: Caculká Hurakán, Chipí Caculká y Raxá Calculká. Estas tres manifestaciones del Hurakán mesoamericano nos recuerda que, según Pané, el cemí Guabancex estaba asistido por otros dos: Guataúba, pregonero y heraldo del viento y la lluvia, y por Coastriquie, quien recogía las aguas para luego derramarlas. Dicho de otro modo, las tres manifestaciones del dios mesoamericano tienen su equivalencia en estos tres cemíes taínos que, a su vez podrían ser manifestación del mencionado Anacacuya, "lucero-del-Centro", equivalente al "Centro-del-Cielo" o Hurakán.

De todos modos, si las deidades relacionadas con un pie o pierna tienen una vinculación directa con el ciclo de la Osa Mayor y los huracanes en el área de Mesoamericana y las Antillas, esta situación parece no ocurrir a nivel suramericano. Allí el héroe mutilado o con una pierna corresponde a la constelación de Orión (Lévi-Strauss, 1968; Magaña y Jara 1982, 1983). Esta transformación simbólica del héroe mutilado o la deidad de la Osa Mayor a la constelación de Orión cuando nos desplazamos de mesoamérica a suramérica, amerita una investigación más profunda. Por el momento, el paralelismo entre algunos mitos taínos con mitos mayas-quiché y suramericanos parece señalar, sin embargo, vestigios de algún tipo de contacto cultural que apenas ha sido estudiado. Cabría pensar que las Antillas fuera un área de encuentro entre dos núcleos culturales o que las islas antillanas pudieron servir de puente de comunicación en época arcaica para grupos migratorios provenientes de suramérica que llegaron hasta la península de Yucatán o la Florida, o quizás oleadas migratorias en sentido contrario, perspectivas estas rozadas por el mismo Lévi-Strauss (1970:51-52) pero que ameritan mucha investigación antes de vislumbrar alguna conclusión.

Por su parte, el análisis de la mitología de los caribes antillanos nos permite pensar que su astronomía primitiva estaba estrechamente identificada con la astronomía de otras tribus caribes suramericanas. Es un hecho histórico bien conocido que los caribes isleños se mantuvieron en contacto con el continente hasta el siglo XVII lo que permite asumir que estos tuvieron "si no las mismas constelaciones que los caribes continentales, al menos algunas en común" (Magaña y Jara, 1983; Magaña, 1984).

Un concenso de los cronistas franceses de las Antillas Menores (Cárdenas Ruíz, 1981) y las investigaciones de este siglo de Taylor (1946), presentan el siguiente panorama de la mitología y la astronomía de los caribes isleños:

Luquo fue el primer hombre y el primer caribe que luego de bajar del cielo, hacer salir de su ombligo y de una incisión en su pierna a los primeros hombres, crear los peces y la mandioca, volvió al cielo. Racumon fue también uno de los primeros caribes que se convirtió en serpiente con cabeza de hombre para luego convertirse en una estrella. Achinaon, Achinnao, Asinnao o Hánnao era una constelación formada por nuestra constelación Altair y otras, era la que producía la lluvia ligera y el gran viento de enero. Las Pléyades, que señalaban el inicio del año, eran conocidas por los hombres como iromobulême, dueña del tiempo caliente, y por las mujeres como sîrik, un nombre equivalente a un cangrejo de tierra que abundaba en la época de lluvias. Bakámo, según Taylor, era una constelación formada en parte por nuestra estrella Antares de Escorpión y por las constelaciones Sagitario y Capricornio, cuyo conjunto constituía el cuerpo de serpiente y la cabeza humana de esta entidad mítica. Epietembo era la constelación del cazador de una sola pierna, correspondiente a Orión. Iaboura, yábura o lukúni yábura, la-canoa-de-la-garza, era una constelación relacionada con la Osa Mayor y la época de tormentas y fuertes lluvias, como ya hemos mencionado. (Figura 7). Quizás esta constelación tenga una identidad con Savacu que, según el cronista La Borde, se convirtió en un gran pájaro y luego, convertido en estrella, era quien enviaba los huracanes, los rayos y truenos que producían grandes lluvias. Otra estrella caribe, Curumon o Kurúmo, fue un ser mítico que al elevarse al cielo producía las grades olas del mar y viraba las canoas.

Puede ser que esta estrella fuera la kwibis o "crayfish" referida por Taylor o la ulihao citada por Breton. Sirio era conocida como malirúbana y Proctón como malirú-bana ápurku, las cuales causaban temporales teniendo los caribes mucho cuidado en no navegar cuando la veían salir pues eran señal de mal augurio. Otras constelaciones eran bihí, representando una vieja mujer; túlulu, un cangrejo de jardín; wanásai, la boa o serpiente. La Vía Láctea se le llamaba, bajo influencia francesa, "chemin la tortue" o el "rastro-de-la-tortuga", pues observando su posición sabían cuando las tortugas vendrían a desovar.

Como podrá apreciarse, las estrellas o constelaciones de los caribes antillanos tenían nombres de animales, aves o entidades míticas y estaban correlacionadas estrechamente con los estados climatológicos que afectaban las Antillas Menores a través del año. Otro detalle interesante que resalta al leer los cronistas franceses es que tanto el Sol como las estrellas servían para la navegación de los caribes en canoas. Bouton (1640) fue el primero en escribir que los indígenas "van de isla en isla y preven con bastante certeza los malos tiempos y tempestades por la inspección del cielo y de los astros, de los cuales tienen conocimientos maravillosos". Años después, Coppier (1645) publicaba que "no faltan de embarcarse en sus canoas, tomando como ruta la estrella del norte, y aún cuando no tienen brújula alguna...ni nada que sirva para medir la altura del sol, no por ello cesan de navegar, tomando su ruta por el curso de algunas estrellas, de la cual no he podido tener conocimiento, a pesar de haberla estudiado". Según este cronista francés el año era contado como doce lunas, "no nombrándolo jamás con otro nombre". Rochefort (1665) señala que "por una larga experiencia unida a la tradición de sus antepasados, han adquirido un grosero conocimiento de varios astros, de donde viene que ellos cuenten los meses por lunas y los años por las Plêyades observando esta constelación". La ocurrencia de eclipses era también de gran trascendencia para los caribes isleños. El eclipse de luna era llamado Laikua noquian ("la consunción de la Luna") y los de Sol Laikua vicu ("la consunción del Sol"). Cuando estos ocurrían, las mujeres y los jóvenes danzaban toda la noche, dando gritos lúgubres y haciendo sonar unas pequeñas piedras dentro de una calabaza. Creían que era Maboya, un espíritu del mal, quien se comía los astros y todo su ritual pretendía alejar el mítico monstruo del camino de la Luna o del Sol para restablecer la normalidad cósmica (Robiou, 1984 b).

A diferencia de los arahuacos y caribes de suramérica, parece que los taínos y caribes isleños desarrollaron grandes conocimientos de navegación estelar, junto a notables destrezas naúticas. De una navegación pluvial y costera en suramérica, estas tribus, a través de migraciones de años, fueron adaptando sus canoas y técnicas a la navegación en pleno mar, lo que les permitió ir poblando las Antillas de este a oeste, en dirección de las corrientes marinas. Para la época del Descubrimiento, no obstante, la comunicación en canoas entre las principales islas era muy común, tal como lo atestiguan los mismos cronistas españoles y la difusión de estilos ceramistas que estudia la arqueología (Robiou, 1985 b).

Además del área de la mitología taína y caribe recopilada por los cronistas españoles y franceses que hasta aquí hemos reseñado, otra área de evidencia de conocimiento de astronomía primitiva en las Antillas son las llamadas plazas ceremoniales construídas principalmente en Cuba, La Española y Puerto Rico muy posiblemente en la época taína. Existen tres tipos básicos de construcciones: los montículos de tierra, los camellones o aceras de tierra y piedras y las de piedras en sí, fuera ya utilizando las piedras en posición horizontal o vertical. Las formas de estas construcciones van desde simples líneas hasta círculos, rectángulos o elipses. El propósito de estas plazas ceremoniales, o *bateyes* no está completamente aclarado. A veces se encuentran en zonas aisladas del área montañosa o en los llanos costeros; otras veces se encuentran, principalmente en Puerto Rico, en grupos o conglomerados de plazas de diversas formas. Algunos han pensado que el llamado "juego de pelota" practicado por los taínos se efectuaba en estas plazas ceremoniales, llegándose a postular una posible influencia del juego de pelota mesoamericano (Alegría, 1983, 1986; García-Goyco, 1984). Desde 1980 hemos investigado las plazas ceremoniales antillanas desde una perspectiva arqueoastronómica y, aunque hoy día la mayoría están destruídas o alteradas por lo que nuestro estudio es solo parcial, creemos que algunas de estas construcciones integraban conocimientos de astronomía nunca antes señalados, (Robiou, 1980, 1981, 1985 c, 1987).

Si la astronomía sirve para ordenar el tiempo, dando lugar a calendarios estelares correlacionados con los ciclos climatológicos y del ecosistema, también debemos considerar la función de la astronomía en el ordenamiento del espacio. Ambos conceptos, tiempo-espacio, están íntimamente unidos pues en un espacio sagrado era el lugar donde el primitivo efectuaba sus ritos con el propósito de recrear, repetir o rememorar lo ocurrido en el tiempo sagrado, el tiempo de la Creación, según señala Eliade (1972). Así, los *bateyes* o plazas ceremoniales antillanas son, pues, la unión del tiempo sagrado y del espacio sagrado dentro de la cosmovisión taína, siendo la astronomía el agente integrador de ambas conceptualizaciones.

El principio de eje del mundo, *axis mundi*, también

parece existir en estas plazas ceremoniales. De acuerdo a la mitología taína, el mito de los cuatro gemelos creadores culmina con la creación de la tortuga-hembra en la espalda de uno de ellos, Deminán, la cual fue engendrada al recibir un esputo o guanguayo lleno de la droga alucinógena cohoba (Piptadenia peregrina) lanzado por el abuelo Bayamanaco, el viejo dios del fuego (Pané, 1974). Este mito taíno, como otros ya señalados, contiene elementos sur y mesoamericanos. Lévi-Strauss (1970, 1972) estudió extensamente una serie de mitos suramericanos llamados la serie de la "mujer grapa". En éstos, el héroe trata afanosamente de librarse de una mujer aferrada a su espalda que se convierte en algún tipo de animal al lograr desprenderla. En el Poul Vuh, Hun-Hunahpú, convertido en calabaza, le lanza un chisguete de saliva a la doncella Ixquic por el cual ésta engendra dos gemelos. En el caso antillano, es un gemelo el que recibe el guanguayo o esputo para engendrar, no en la barriga sino en la espalda como en los mitos suramericanos, una tortuga-hembra.

La procreación de la mujer, o tortuga-hembra, luego de la existencia de cuatro héroes gemelos, recuerda el mito azteca cuando en "Las-cinco-serpientes-de-nubes", los cuatro hombres corresponden a los cuatro puntos cardinales y el quinto, la mujer, corresponde al centro o a la dirección arriba-abajo, apareciendo más tarde en mesoamérica en el juego de pelota que es, según Krickeberg (1971), símbolo de esa dirección. De ser así, los cuatro gemelos taínos, símbolos de las cuatro direcciones, también podrían ser símbolos de las cuatro fechas principales del ciclo solar (dos solsticios y dos equinoccios) que corresponden a los cuatro períodos de tiempo casi iguales en los que se divide el año solar y a sus respectivas cuatro posiciones del Sol en el horizonte. De este modo, los cuatro gemelos taínos podrían tener una estrecha relación con el origen del juego de pelota y las plazas ceremoniales antillanas y la tortuga-hembra vendría a ser símbolo de la dirección arriba-abajo, de la unión entre lo divino y lo humano, unión que se recrearía a través de los ritos en un espacio y durante un tiempo sagrado (Robiou 1983 b, 1984 a, 1985 a).

La plaza elipsoidal de Chacuey, en Dajabón, República Dominicana, está formada por un camellón de piedras con un ancho de casi 5 metros, interrumpido en dos especies de entradas. (Figura 8). El eje de estas entradas o portales es perpendicular al eje principal de la elipse, el cual tiene una orientación de 28 ^{o} 30' y mide unos 243 metros. El eje transversal alcanza unos 147 metros, teniendo la plaza un área de unos 29,000 metros cuadrados. De la plaza elipsoidal salen dos calzadas paralelas que terminan en el rio Chacuey, donde hay un complejo de petroglifos únicos en las Antillas (Boyrie Moya, 1955).

De acuerdo a los planos elaborados en 1955 y nuestra investigación de campo en 1980 creemos que el eje de los portales de la plaza ceremonial de Chacuey está orientado hacia la salida del Sol en el solsticio de invierno. La dirección de las calzadas paralelas que bajan al río están orientadas hacia los equinoccios, mientras el montículo de piedras localizado en el área noreste de la plaza, de unos 24 x 40 metros, cuando se observa desde el centro el portal occidental, está en dirección de tres fechas astronómicas significativas: la salida del Sol en el solsticio de verano y los días de su paso por el cénit y la salida heliacal de las Pléyades. La plaza elipsoidal de Chacuey, la mayor en su tipo, demuestra al grado de conocimientos de geometría, astronomía y organización social que llegaron sus constructores taínos probablemente años antes del Descubrimiento. Sus dimensiones no responden al área requerida para practicar el llamado juego de pelota descrito por los cronistas españoles, por lo que podemos pensar que el sitio era un lugar de peregrinación y reunión para efectuar determinados ritos, quizás de implicaciones astronómicas, en ciertas fechas del año.

Otra plaza ceremonial que también parece contener orientaciones astronómicas es el llamado Círculo de los Indios, en San Juan de la Maguana, República Dominicana. (Figura 9). La construcción circular tiene un diámetro de unos 225 metros y está formada por una doble hilera de piedras. El monolito central de 1.7 metros de altura muy posiblemente podía servir de mira para una alineación hacia la salida del Sol en el solsticio de invierno con la Loma del Agua. Ese monolito central, quizás símbolo de una deidad solar, podría también servir para marcar los días del paso del Sol por el cénit (15 de mayo y 29 de julio), días en los cuales su sombra sería más corta. Sabemos que varias culturas de América tropical y algunas tribus del Orinoco en particular, observaban los días del paso del Sol por el cénit con el propósito de celebrar ritos agrícolas. Aunque no tenemos evidencia de los taínos, sí sabemos que los caribes isleños conocían y observaban estos días (Taylor, 1946). En Puerto Rico, isla donde se reportan más de 40 plazas regularmente rectangulares pero de diversos tamaños, García Goyco (1984) informa de orientaciones hacia los equinoccios en las plazas principales de los centros ceremoniales de Caguana, en Utuado y Tibes, en Ponce.

Conclusiones

Los cronistas españoles nos ofrecen limitadas referencias sobre los conocimientos de astronomía que tenía la desaparecida cultura taína de las Antillas Mayores. Tenemos evidencias de la importancia del Sol, la Luna y los eclipses en la cosmovisión taína, así como sus efectos climatológicos para una sociedad agrícola, pero no poseemos referencias directas sobre las estrellas y otros astros. Sin embargo, el análisis de su mitología nos permite entresacar tres mitos posiblemente asociados a Orión (época de sequía - pesca), las Pléyades (época de lluvia - cultivo) y la Osa Mayor (época de huracanes-navegación).

Creemos haber encontrado un paralelismo simbólico entre la deidad maya-quiché Hurakán y la taína Anacacuya, ambas relacionadas a la Osa Mayor y su ciclo alrededor de la Polar, como un posible calendario que señalaría la época de huracanes en las Antillas y Yucatán. Las evidencias de alineaciones astronómicas en algunas plazas ceremoniales antillanas donde se efectuaba el batey o juego de pelota, hacen pensar que esta era una actividad con simbología cosmogónica, al igual que ocurría en mesoamérica. Estas similitudes requieren una amplia y profunda investigación por parte de especialistas.

Los cronistas franceses de las Antillas Menores nos ofrecen un panorama más preciso de la importancia de la astronomía para los caribes isleños, la cual vinculaban intimamente a su mitología, la navegación en canoas y los estados climatológicos. Sin embargo, las diferencias o grados de desarrollo en los conocimientos sobre astronomía entre taínos y caribes insulares no puede establecerse en base a la cantidad de información recopilada por los cronistas.

En las Antillas, sin dudas, la astronomía era parte integral de la cosmovisión de sus primitivos habitantes que lamentablemente no lograron sobrevivir al impacto de la conquista y la colonización europea.

- - - -

Bibliografía

Alegría, Ricardo E.

1981 Prólogo a *Crónicas Francesas de los Indios Caribes*. San Juan: Editorial de la Universidad de Puerto Rico.

1983 *Ball Courts and Ceremonial Plazas in the West Indies*. New Haven: Yale University Publications in Anthropology, No. 79.

Arrom, José Juan

1974 *Relación Acerca de las Antigüedades de los Indios, de fray Ramón Pané (Versión)*. México: Siglo XXI.

Aveni, Anthony F. & Gary Urton, eds.

1982 *Etnoastronomy and Archaeoastronomy in the American Tropics*. New York: The New York Academy of Science, Vol. 385.

Boyrie Moya, Emile de

1955 *Monumento Megalítico y Petroglifos de Chacuey, R.D.* Ciudad Trujillo: Universidad de Santo Domingo.

Colón, Cristóbal

1961 *Diario. Los Cuatro viajes del Almirante y su testamento*. Madrid: Espasa-Calpe, S.A.

Cárdenas Ruíz, Manuel

1981 *Crónicas Francesas de los Indios Caribes*. San Juan: Editorial de la Universidad de Puerto Rico.

Eliade, Mircea

1972 *Tratado de Historia de las Religiones*. México: ERA

García Goyco, Osvaldo

1984 *Influencias Maya y Aztecas en las Antillas Mayores*. San Juan: Ediciones Xibalbay.

Krickeberg, Walter

1971 *Mitos y Leyendas de los Aztecas, Incas, Mayas y Muicas.* México: Fondo de Cultura Económica.

Las Casas, Bartolomé de

1951 *Historia General de las Indias.* 3 vols. México: Fondo de Cultura Económica.

Lehmann-Nitsche, R.

1924 "La Constelación de la Osa Mayor y su concepto como huracán o dios de la tormenta en la esfera del Mar Caribe". *Revista del Museo de la Plata.* Tomo *XXVIII: 103-145.* Buenos Aires.

Lévi-Strauss, Claude

1968 *Lo Crudo y lo Cocido.* México: Fondo de Cultura.

1970 *El Origen de las maneras de mesa.* México: Siglo XXI.

1972 *De la miel a las cenizas.* México: Fondo de Cultura.

Magaña, Edmundo

1984 "Carib Tribal Astronomy". *Social Science Information.* No. 23 (2): 341-368. SAGE, London.

Magaña E. & F. Jara

1982 "The Cribe Sky" *Journal de la Société des Américanistes.* 68: 105-132. Musée de L'Homme, Paris.

1983 Astronomy of the Coastal Caribs of Surinam". *L'Homme*, XXIII (I): 111-133.

Mártir de Anglería, Pedro

1964 *Décadas del Nuevo Mundo.* 2 vols. México: Porrúa e hijos.

Ortíz Fernando

1947 *El Huracán, su mitología y sus símbolos.* México: Fondo de Cultura. (1984).

Oviedo, Gonzalo Fernández de

1959 *Historia General y Natural de las Indias*. 4 vols. Madrid: Biblioteca de Autores Españoles.

Pané, fray Ramón

1974 *Relación Acerca de las Antigüedades de los Indios*. México: Siglo XXI.

Robiou Lamarche, Sebastián

1980 "Arqueoastronomía en Chacuey". *Listín Diario*, 27 de diciembre: 13-15. Santo Domingo, R.D.

1981 El Corral de los Astros", *Listín Diario*, 11 de junio: 8-9. Santo Domingo, R.D.

1983 a "Del Mito al Tiempo Sagrado: un posible calendario Agrícola- Ceremonial taíno". *Boletín* Museo del Hombre Dominicano, XI-18: 117-140. Santo Domingo, R.D.

b. "Astronomía en la Mitología Taína", *Listín Diario*, 13 de agosto: 14-15. Santo Domingo R.D.

1984 a "Astronomy in Taíno Mythology". *Archaeoastronomy*. The Journal of the Center of Archaeoastronomy. Vol VII No. 1-4: 110-115, Maryland, USA.

b. "Eclipses en la Prehistoria Antillana", *Listín Diario*, 15 de diciembre: 4-6. Santo Domingo, R.D.

1985 a "Ida y Vuelta a Guanín: un ensayo sobre Cosmovisión Taína". En *Myths and the Imaginary in the New World* (Magana & Mason, eds., 1986). Amsterdam: CEDLA.

b "La Navegación Indígena Antillana". *Isla Abierta*, suplemento de *Hoy*, 8 de junio: 4-7. Santo Domingo, R.D.

c "Panorama de la Arqueoastronomía en las Antillas". *Primer Congreso Nacional de Arqueología*, Museo del Hombre Dominicano, septiembre. Santo Domingo, R.D.

1987 a "Posibles Símbolos Astronómicos- Meteorológicos en el Arte Rupestre Antillano". *VIII Simposium Internacional de Arte Rupestre Americano*.Museo del Hombre Dominicano,

8-13 de junio. Santo Domingo, R.D. p.p. 405-429.

b "Panorama de la Astronomía Indígena en las Antillas" _Boletín_ Museo del Hombre Dominicano, XIV-20: 83-95. Santo Domingo, R.D.

Roth, Walter E.

1915 _An Inquiry into the Animism and Folk-lore of the Guiana Indias._ Bureau of American Ethnology. XXX Annual Report (1908-09). Washington: Smithsonian Inst.

Sued Badillo, Jalil

1978 _Los Caribes: realidad o fábula._ San Juan: Ed. Antillana.

Taylor, Douglas

1946 "Notes on the Star Lore of the Caribbees". _American Anthropologist._ Vol. 48: 215-222.

- - - -

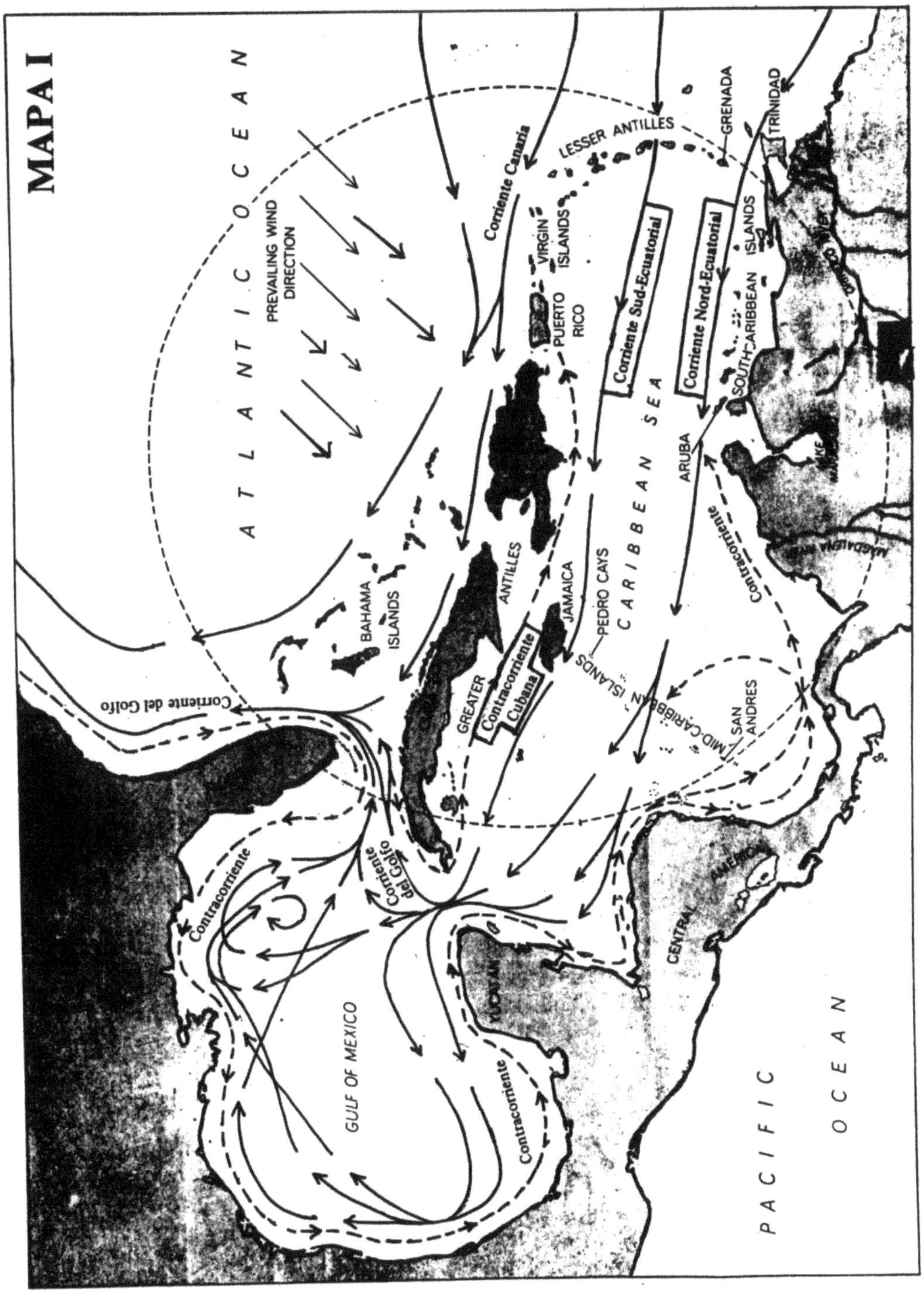

Fig. 1 Las Antillas

Fig. 2 Colón e indios jamaiquinos durante eclipse lunar, 29 de febrero de 1504

Fig. 3 Idolo gemelo taino, posible representación de Máhoru y Boinayel

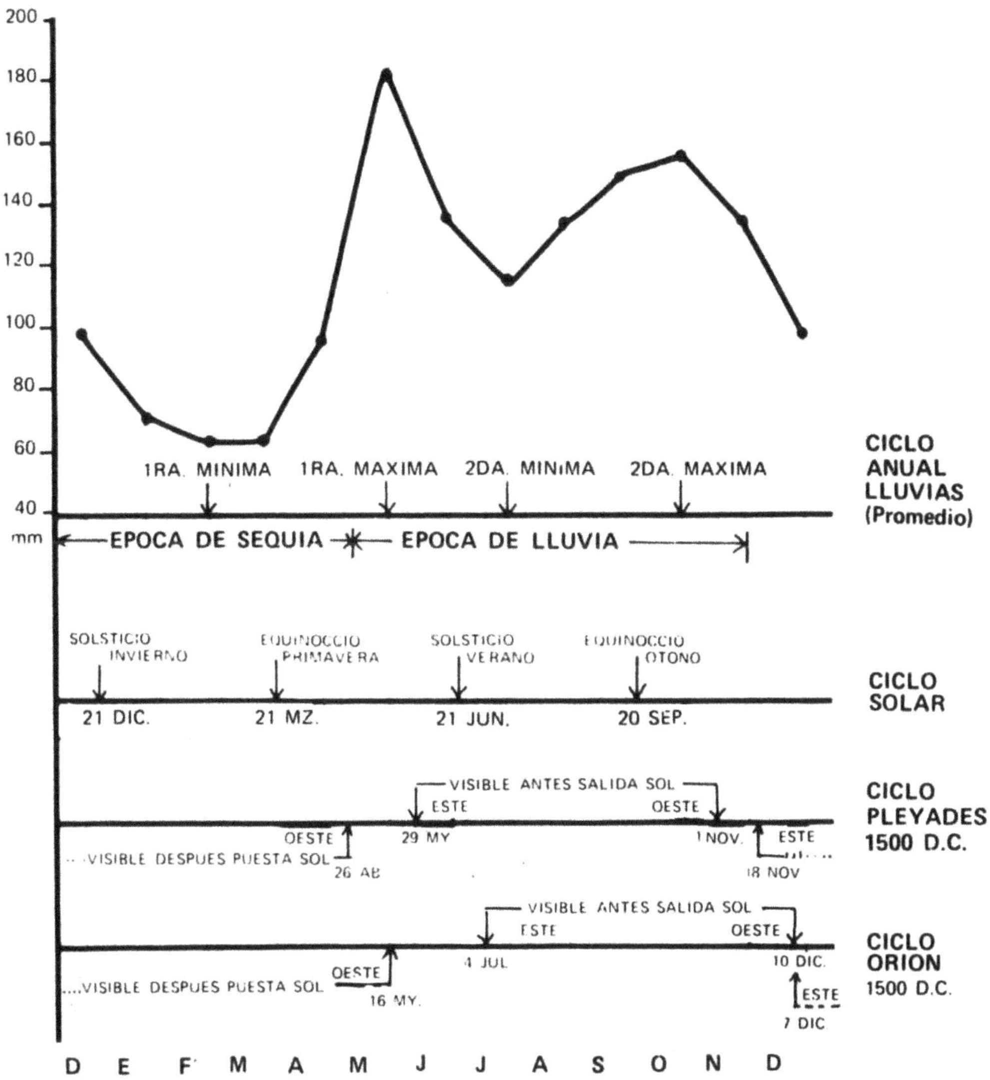

Fig. 4. Correlación entre el ciclo anual de lluvias en las Antillas, el ciclo solar y los ciclos de las Pléyades y Orión (Robiou, 1983).

Fig. 4 Correlación entre el ciclo anual de lluvias en las Antillas, el ciclo solar y los ciclos de las Pléyades y Orión (Robiou, 1983)

Fig. 5 Sello o pintadera de barro taino con motivo batriforme. Universidad de Puerto Rico

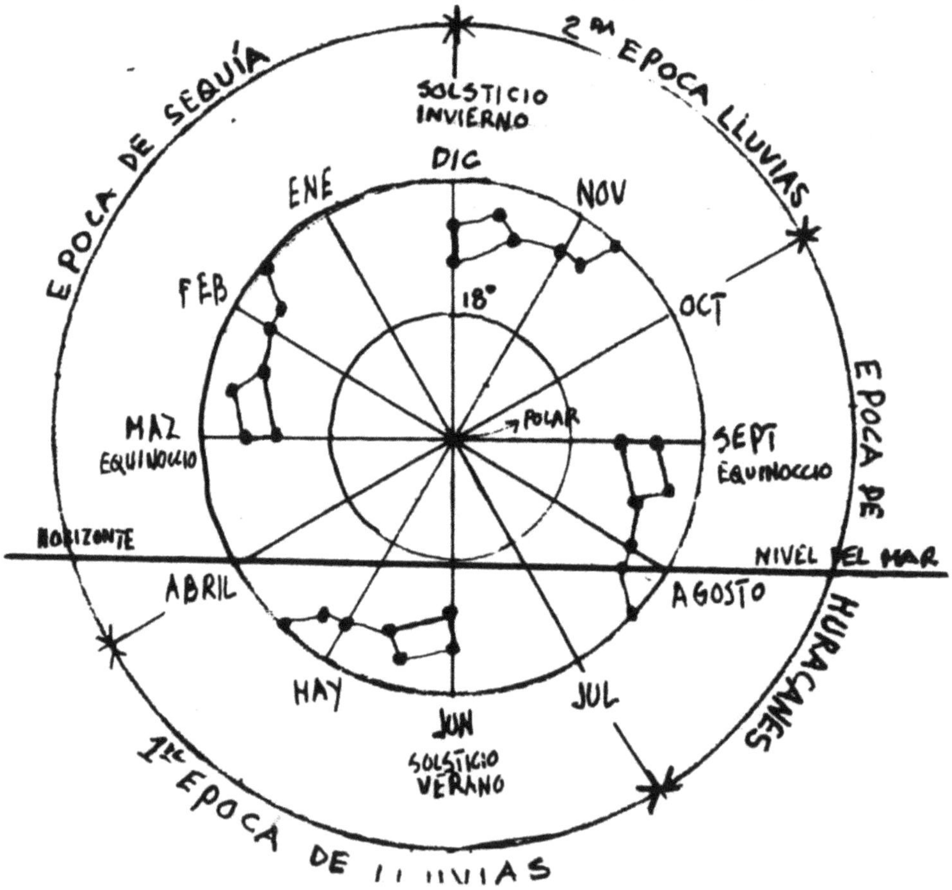

Fig. 6 Ciclo anual de la Osa Mayor alrededor de la Polar, la cual está a 18° sobre el horizonte, equivalente a la latitud promedio de las Antillas

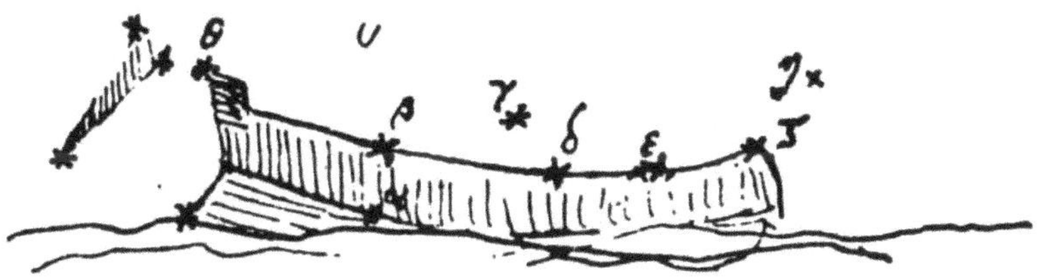

LUKÚNĪ YÁBURA

Fig. 7 'Lukúni-Yabura'-constelacion 'la-canoa-de-la-garza', en parte la Osa Mayor (Taylor, 1946).

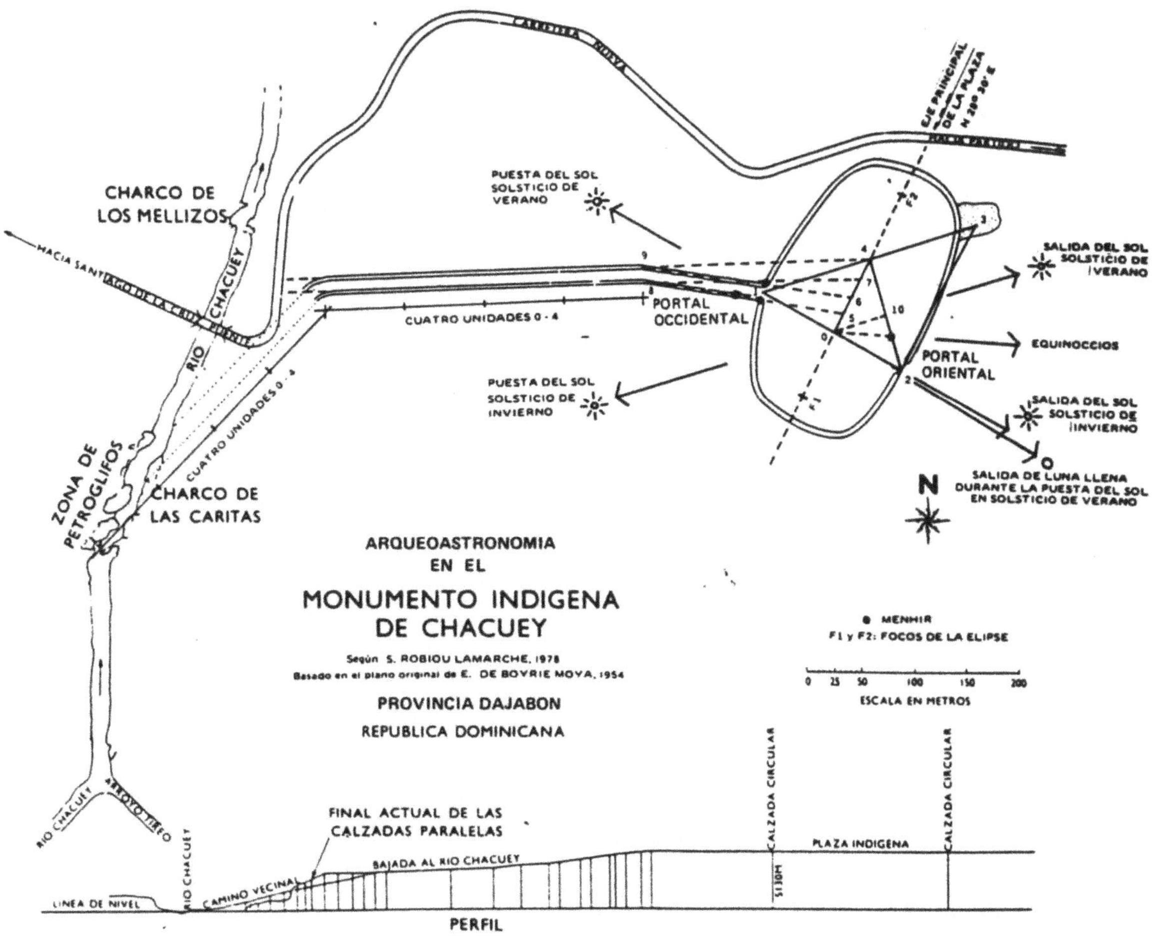

Fig. 8 Arqueoastronomia en el Monumento Indigena de Chacney

R. Tom Zuidema

The pillars of Cuzco: Which two dates of sunset did they define?

Introduction

In an earlier analysis of the Incaic observations of sunset in between a set of four pillars --two inner ones and two outer ones-- erected on the nearby horizon of their capital Cuzco, Aveni (1980, 1981) and I (1980, 1981) suggested that the sunset in between the central pillars might have marked the antizenith sunset (18 August, 25 April); this observation being made in the opposite direction of where, at this latitude (-13.5), the sun rises when going through zenith (30 October, 13 February). One reason for making this suggestion was that an Anonymous author (p. 151), probably writing in ± 1560-1570, states that the observation was made in the month of August. At that time he would have used the Julian calendar (valid until 1584) allowing for a possible date of the particular sunset anywhere in between 10 August and 10 September (Gregorian calendar). Moreover, the observation was made from a central location in the plaza of Cuzco, the Ushnu, and he (p. 158) and Molina (p. 30,31,46) stress its ritual importance only for the month of September during a new and the next full moon. This information, combined with that of other chroniclers on the pillars and on the times of planting and of harvest (Betanzos, Sarmiento and Cobo) make it more plausible that the central observation was made for a date close to the beginning of September.

In this paper I will not enter into a further discussion of these ethnohistorical data, nor into that of the exact locations of the Ushnu and the pillars. I will use other information that can help us to discover the astronomical reason for making the observation; a reason that the Spanish chroniclers never mention explicitly as it did not seem to conform to either of the dates of the solstices, the equinoxes, the passages of the sun through zenith or even the antizenith sunsets.

My conclusions will be that:
1) the observation was made for lunar reasons that help to define the first full moon on or after the September equinox;
2) the equinoxes, like the solstices, were not defined by an exact day, but by a period of eight days; and
3) the September equinox observation was related to the South in the cosmological system of the Incas in Cuzco unifying space and time.

Part of my analysis is based on information not from Cuzco. We have no data on sets of pillars from other Incaic towns, although they may have existed. But we can understand the reason why they were erected in Cuzco in a more general Andean context and we can analyze a corpus of Andean data, related to the observation of the moon, which in no way represents a kind of automatic system considering all moons in the year equally. Some lunar observations were privileged and each depended on solar observations immediately related to the particular lunar event.

While this analysis is made independently from the calendar as based on the _ceque_ system, it will serve as a check upon the construction of the latter. The ceque calendar counted 328 toponyms in a calendrical order of so many days, codified in a system of 41 _ceques_ or 42 directions (2 directions being counted together as one ceque), seen from the central temple of the Sun in Cuzco. This analysis will support the conclusion that the _huacas_ (the places to which the toponyms applied) on the ceques going South were counted around the September equinox and it will provide an argument why this is so.

My argument develops in four sections. In the first, I consider the systematic pattern behind the attention given to only certain moons in the year as described by the chroniclers for the calendar of Cuzco. In the second, I analyze the part of a treatise on Inca astronomy and the calendar where the indigenous chronicler Felipe Guaman Poma de Ayala stresses the importance of the date of 29 August for the solar-lunar relationships around the September equinox. In the third, I confront my conclusions with those based on the ceque calendar. Finally, in the fourth section, I discuss the importance of the South direction for determining the September equinox observation and I stress the further concordance of this result with the calendar as based on the ceque system. This argument again departs from Guaman Poma's treatise but now from an other part of it.

Solar observations and lunar celebrations.

Considering that in the western calendar all movable feasts in a given year --like Easter, Pentecost, Eucharist and Carnaval-- follow in a regular lunar sequence upon the first, calculated from the vernal equinox and not taking into account any other solar observations, we should observe in the case of the Incas that such a system was not applied. Although they also celebrated only specific moons in the year, they, in contrast, selected each lunar observation within a specific period of the year as defined by two succeeding solar observations. Moreover, more attention was payed to certain moons during the time from sowing to harvest, when a female interest dominated, than to those during the other time of the year. Not only the specific moons, but also the system of months was defined by the system of multiple solar observations. Therefore, I will mention these first (Zuidema 1981, 1982 a, 1982 b).

As far as we can assess our information, the Incas made six kinds of solar observations, each from their own and independent observatory in or around Cuzco. Thus they generated ten precise dates as recorded in their annual rituals. Five of these observations do not offer any problems, as our information, obtained from the chronicles and from fieldwork in Cuzco, is unambiguous. I mention these observations first, together with the dates that they generated and the events that were celebrated, before I discuss the problem of the sixth observation which is the subject of this article.

1) Sunset during the June solstice (JSSS) was observed from a temple North of Cuzco and was celebrated in a feast called Inti raymi (the feast of the Sun") during the full moon on, or just before or after, the event (Molina p. 25,27).

2) Sunrise during the December solstice (DSSR) was observed from a temple South of Cuzco and fell in between two Incaic months both called Capac raymi (Capac raymi "royal feast" and Capac raymi Camay quilla "royal feast, the moon (of) Camay"). While no reference is made to the moon in the rituals of the first month, the more important ones of the second month occurred during full moon and the five days after (Molina p. 61-6 ; Zuidema and Urton 1976). The December solstice itself appears to have been associated with a new moon around this event.

3) Two dates, about one month before and after the June solstice respectively, were defined by way of an alignment of the temple of the Sun in Cuzco towards sunrise. Thus the precise date of the June solstice could be calculated. But the observation was said to be made in order to calculate the earliest new moon of the month of Inti raymi. Such a calculation allows us, and probably allowed the Incas, to define the fixed period of a month around the June solstice that contained all the possible full moons of Inti raymi. The reverse of this alignment towards sunset could have been used to define exactly the date of the December solstice. But here we miss all explicit information.

4) Two dates, 30 October and 13 February, when the Sun goes through Zenith in Cuzco, were defined calendrically by observing sunrise on those days from a mountain, called Picchu, just West of Cuzco. The whole period of 107 days between those two dates was important as then the noble initiates were brought into contact with the ancestors. It was opened with rituals just after 30 October and closed before 14 February (Zuidema, in press b). Combining this information with that on the moon, it will be clear that among the four possible full moons that could occur within the 107 day period only two were important for celebration: one in between the first date of zenith passage (30 October) and the December solstice (21/22 December) and the other in between this solstice and the second date of zenith passage (13 February).

5) By reversing the observation of the two zenith sunrises --that is, looking towards mt Picchu--, two other dates were generated, half-a-year apart from the first two, that, for convenience sake, I call those of antizenith sunset (25 April, 18 August). The direction of zenith sunrise was also used for a procession in the reversed antizenith direction, somewhere in the month of April (10 April - 10 May, Julian calendar) when the antizenith sunset occurred (Zuidema, in press a). But nowhere in the chronicles do we find any reference to the concept of antizenith. No ritual is attached to the second date of antizenith

sunset, 18 August. We lack all information about the moon in relation to both dates.

6) The sixth observation, that of sunset in between the set of four pillars --two smaller, inner ones and two larger, outer ones--, presents us with a different kind of problem as no astronomical reason is given for its observation, only a calendrical one. The pillars were erected on the slope of mountain Picchu, the mountain that was used also for observing the antizenith sunset. Technically, the two kinds of observation were quite different from each other. One was carried out from a distance of ± 30 kms and in this case the sun's diameter was larger than the width of the mountain; the other was made from a distance of 2 kms, where the Ushnu stood in the plaza of Cuzco, and now the pillars were used occupying only part of the mountain. There is no reason to assume off-hand that the two kinds of observation served one and the same calendrical purpose. 1/ We know the general location of the pillars and of the Ushnu Aveni 1981; Zuidema 1980, 1981). The passage of the Sun through the system of pillars can be adjusted just as well to a month around the antizenith sunset, 18 August, as to one around the beginning of September, some two weeks later, involving only a minor shift in the locations of the pillars and the Ushnu. The calendrical information agrees, however, better with the second date. Lunar evidence clarifies the astronomical and calendrical motivation why the Incas erected the pillars. Before I proceed with these specific lunar arguments, let me mention first in more general terms the problem we are confronted with.

Except for the one month around the June solstice, there was a distribution of periods in the year, defined by the solar observations just mentioned, that show a pattern of mirror symmetry around the temporal axis of the solstices. The four successive periods of most interest at the moment are:
1) from the solar observation of sunset in between the central pillars, probably around the beginning of September, to the first zenith passage, 30 October;
2) from there to the December solstice;
3) from there to the second zenith passage, 13 February; and
4) from there to the second sunset in between the central pillars, probably around 10 April.
For the sake of convenience I identify these periods as 1, 2, 3 and 4. The lunar evidence allows us to argue for a similar length of the four periods. Periods 2 and 3 are each other's mirror image in terms of solar movement along the horizon and so are periods 1 and 4. We have one kind of lunar information on period 2 and another on period 3. Although the chroniclers do not express these two kinds of information in exactly the same terms, they can be taken as referring to the same kind of lunar observation and we can make the inference that, not only the solar information represents both periods in a mirror symmetry, but the lunar information does so too. Thus the lunar information on period 2 helps to clarify that on period 3 and vice versa, the same being true for periods 1 and 4.

Coming now to the lunar data, the Anonymous (p. 159) states

that the two Incaic months corresponding to the Spanish ones of October and November were "celebrated in one way", "because the Incas would knight their youths during the full moon of these two lunar months". He makes a similar statement (p. 156) about the Incaic months corresponding to the Spanish ones of March and April. Here he adds, however, the statement that these two lunar months, each "going from one conjunction to another", were taken together "because the moon of March always reaches the one of April". Apparently, a similar reason existed for making the first statement about the one moon of two months. Both statements can mean only that the Incaic months themselves were not movable; they had a fixed position in the year. Otherwise, the moon never would have been able to pass the boundary of its corresponding month. They also mean that each of these two double-months necessarily contained less days than two synodic months; if they were equal to them, it could happen that each of the Incaic months contained a movable synodic month that did not reach the other. We also know that important lunar feasts were celebrated in the Incaic months corresponding to September and January (Molina, Cobo), but in no other month; not even in the month of Inti raymi when a lunar observation was made in relation to the June solstice. Periods 2 and 3, as defined by the zenith passages and the December solstice, each measured a period of 53.5 days. They do conform, therefore, well with the statement of the Anonymous about the two Incaic double-months, these being shorter than two synodic months. My first conclusion, therefore, is that the two double-months mentioned by the Anonymous correspond to periods 2 and 4, as defined by the solar observations. My second conclusion is that the reason for mentioning the one moon of two months in period 2 was similar to the reason why in period 3 only one important lunar celebration is mentioned and not two. My third conclusion is that the argument, given in the second conclusion, becomes valid also for periods 1 and 4. The Incas celebrated the full moon corresponding to September following the sunset in between the central pillars. Therefore, periods 1 and 4 --those from pillar passage to zenith passage and vice versa-- also should be some days shorter than 59 days. The central pillars probably indicated dates somewhere in the beginning of September and around 10 April, the dates being less than 2 x 59 days before and after the December solstice.

Fitting the double-months as mentioned by the Anonymous into the sequence of the four periods, I have to refer briefly to the names of the months. He identifies the first double-month with the Spanish months of October and November; but their Quechua names --Chaupicuzqui, also called Cantarayquilla, and Raymiquilla, respectively-- and the description of their rituals indicate that we are dealing with the two months before the December solstice. We can identify them with period 2, going from first zenith passage (30 October) to December solstice (53.5 days). The correlation of the other double-month to Spanish months requires more discussion. All the Spanish chroniclers tried, in one way or another, to fit the Incaic system of monthly divisions into the Procustean bed of their own system of 12 months. However, it is clear from the Incaic system of solar observations, and from the fact that only one month Inca raymi was recognized around the June solstice against two months Capac raymi around the December solstice, that a system of 13 monthly divisions fits better into their concept of the year. The Anonymous, or his informant, comes with a solution to the problem of month names that no other author offers in quite the same way. He first intercalates a month called Haocaycusqui --the name that other chroniclers use for the month of Inti raymi-- in between the two months of Ayriuaquilla and Aymorayquilla, otherwise always mentioned one after the other, and then he identifies the first two of these as the double-month of March-April. Given these kinds of shifts in months or month names, an argument about their "right and original" order, when offered in isolation from other names and lists of month-names, can go in any direction. I suggest that his information on Ayriuaquilla and Haocaycuzqui as a double-month corresponds best to the fourth Incaic period and to the months that most other chroniclers call Pachapocoy and Ayrihua. 2/ In this way we can apply the argument as made for periods 2 and 3 also to periods 1 and 4. The extensive description of lunar celebrations in the month corresponding to September, but not in October, matches the reference to "the moon always reaching the other month" in March-April. September-October would coincide with period 1, going from first pillar passage to first zenith passage, and March-April with period 4, going from second zenith passage to second pillar passage.

Guaman Poma and the 29th of August.

Another lunar argument for locating the central pillars in time is included in a particularly important but abstruse text of some two pages where the indigenous chronicler Felipe Guaman Poma de Ayala (p. 883-5(897-9)) explains the astronomical importance of the Sun, the Moon and the stars in the Inca calendar. He also mentions in this context a ritual event that had occurred some years before he heard about it and that elucidates the ritual importance of the two months following the September sunset in between the central pillars. His treatise reveals, without mentioning the pillars, an astronomical and calendrical reason for their existence.

Guaman Poma begins his description of the Inca months paying special attention to the day of the death of St John, 29 August. While the date celebrating his birth, 24 June, is better known, medieval legends about the births and deaths of him and of Christ, who were the only Saints having both of them celebrated, clarify the picture (Attwater, Voragine). We can understand the day of his death, 29 August, against the background of the three other ones: the day of the birth of Christ as associated with the December solstice; the one of the birth of St John with the June solstice; and the one of the death of Christ with the first full moon after the March equinox. The two dates of death seem to imply also a significance in relation to the two dates of birth. When Mary became pregnant she visited St Elisabeth, the mother of St John, who by then was six months pregnant, and she stayed with her until the birth of St John. In other words, Christ died at about the same time of the year when Mary had become pregnant and the same was true for St John and St Elizabeth. The two pregnancy cycles as well as the two dates of death integrate into the calendrical divisions of the year by way of the solstices and lunar dates near the equinoxes. The reason why a lunar calculation was chosen for the death of Christ may have been due to the lunar date chosen for his conception and the date of the 29th of August may have been of a similar importance in relation to St John. If so, it would mean that 29 August was chosen as a first possible date of a full moon related to the September equinox.

Guaman Poma implies that after 29 August the days become longer than the nights. His use of the Spanish language is completely confusing and needs a full quotation of the relevant phrases:

> And thus (is) the run of the sun's wheel in summer and winter, from the month that begins with January. The philosopher says that one day he (the sun) seats himself in his seat and rules and takes possession from that principal degree.
> And in the same way in the month of August, the day of St John the Baptist, he seats himself in the other seat; in the first seat of his arrival, in the second seat he does not move from there. In his first principal day he rests and rules from that degree. On the third day he begins to move and makes himself ready for his travel but only very little; it is therefore said that he makes himself ready.
> And from that degree he begins moving every day, without resting, about half an hour, towards the left, facing the Northern Sea (the Atlantic), the six months from the first month of January.

He then mentions, however, not six months as expected, but seven, including July. He continues:

> With (or "after"?) this month of August he (the sun) begins again, from his principal seat to the second principal seat, as he has these two seats much in his power and as each month has its own seat in each degree of heaven; the sun as well as the moon who follows as (his) wife and as queen of the stars. She follows the man as he continues indicating (the degrees) and (as he functions as a) clock of the months and of the year.

Guaman Poma then enumerates the months from August to December. Even if at one moment he had referred to six months beginning with January, he clearly starts the return trip of the sun not with the month of July as the month after the June solstice, but with the month of August, and even here he chooses one of its last days, 29 August, for the beginning of the return. Thus, he really takes September as the first month of a return period of only four months. One reason why intuitively he might not have chosen for July is his awareness that the Incaic month of June fell around the date of the June solstice and not before, and thus that July could not begin with this event. But his references to the day of the death of St John, 29 August, and to the moon who as a queen follows the sun, makes us aware that he really meant the Incaic month corresponding to September, Coya raymi, the "month of the Queen", as the first month of return.

The next part of these pages on astronomy and the calendar seems to be a discussion of the importance of Coya raymi in terms of planting, the moon, the equinox --which theme is the principal interest of this section--, the stars, and the rituals that I will discuss in the fourth section. First, Guaman Poma says:

> And in order to know the hours and minutes, the astrologuer says that he looks early in the morning to the rays of the sun and that one has to look at the clear (sky) through a window; the clear (point) of a mountain where (the sun) rises or sets, and that thus he does not misjudge the point and road of the sun. And there they see at what time one has to sow the early and late crops, and for planting and for shearing the wool of the sheep (llamas), pacos (alpacas) and all the food and fruit that they eat and have to leave eating because of the illnesses in their months giving bad effects and illnesses and dangers.

He is most concerned with sowing and illnesses, two concerns that point to his interest in September as the principal month of sowing and of making precise astronomical observations. We now come to the most important and most confusing part of his text:

> And they say that from the month of January the day is very long and the night short and from August the day short and the night long. And they say that the moon is one degree low in heaven, the sun a very high degree and that she is the wife and the mistress of the sun. And they imagine the sun as having a beard like men and they want to shear the beard for harvest, Yntip cha[c]ranta suncayta tirasac ("I will tear out the hair of his beard, the crop of the sun"). And in the same way they said to the moon Coya Raymi, to the sun Ynti Raymi. During the eclipse of the moon they shouted to her Quilla mama, ama uncuycho, ama uanuychu, cozanchicca olconchicca macacoctacmi anyacoctacmi ("Mother moon, don't become sick, don't die; our husbands, our men, who beat us, who quarrel with us").

Guaman Poma mentions two feasts: Inti raymi, around the June solstice, and Coya raymi, in September. In June, he says, the sun looses its beard; that is, it looses its rays because of its weakness in winter. He relates lunar eclipses to September. And though he says that husbands quarrel with their wives during eclipses, the point he makes is that in September they should not quarrel, this in order to insure a good crop (Molina p. 35; Cobo p. 218; Murua p. 108). The reference to eclipses seems to be less to real ones than to an event recurring every year! It is in this context that the first part of the quotation becomes important. Irrespective of his faulty way of saying, Guaman Poma must have meant with his reference to January, not that days were long after

January, but that they were longer than the nights around the December solstice: that is, from the September equinox to the March equinox. In the same way, reading that days were short and nights were long after August, must have meant that such was the case upto September.

Finally, with reference to the relative positions that Guaman Poma mentions of the sun and the moon in the sky, we realize that it are the sun and the full moon who change their higher and lower courses around the September equinox. While before the sun had followed a lower course than the full moon, now he follows a higher one. As a lunar eclipse can only occur during a full moon, women would have feared most the full moon in September, the one around the equinox, when, because of their interest in sowing and planting, they did not want her to become sick and die and they did not want their husbands to beat them and to quarrel with them. In this month, warriors chased out of Cuzco the evil and illnesses that affected equally the health of the moon, the crops, animals and men (see section 4). This male ritual was carried out during the new moon of September and therefore done in order to prevent what might happen to the full moon after, around the September equinox itself. With an average full moon on 21 September, the preceding new moon would fall on 6 or 7 September.

There is a discrepancy of about 8 or 9 days between the two dates of 29 August and of 6 or 7 September, both culled from Guaman Poma's text. Before I discuss that discrepancy in the third section, let me finish with the part of his text where he refers to the stars:

> These shoutings (to the eclipse) were done by the women and the philosophers said that they know that in the stars are men and women and sheep (llamas) with their young and partridges and a hunter and a herder, a grinding stone, a lion, a deer. And they know from the comets what is going to happen as a good or bad sign: cuyllorcona (bright stars), Chasca Cuyllor ("star of the bright rays": Venus as morning star), Oncoy Cuyllor ("star of illness": Pleiades), Caza Cuyllor (Aymara or Quechua?: ?), Pacari Cuyllor ("morning star": Venus), Uara Uara (Aymara: "star" (not a plural term)).

Guaman Poma refers with his Quechua names first to four bright stars or planets and then to stars in general. Although I suspect that he gives us here some systematic information, it is still too inconclusive to be of much help for understanding the rest of his description. He contrasts them, however, to constellations, identified by their Spanish names, where we can guess about his intent. The Llama with her Young and the Partridge (really a tinamou; Urton 1981 p. 102) are dark cloud constellations in the Milky Way: the first includes the stars Alpha and Beta Centauri as the Eyes of the Mother Llama and the second is the Coalsack in the Southern Cross. Zuidema and Urton discussed these constellations and their calendrical importance for the months of September and October, when, as quasi-circumpolar stars and in succession, they disappear after sunset reappearing before sunrise (Zuidema and

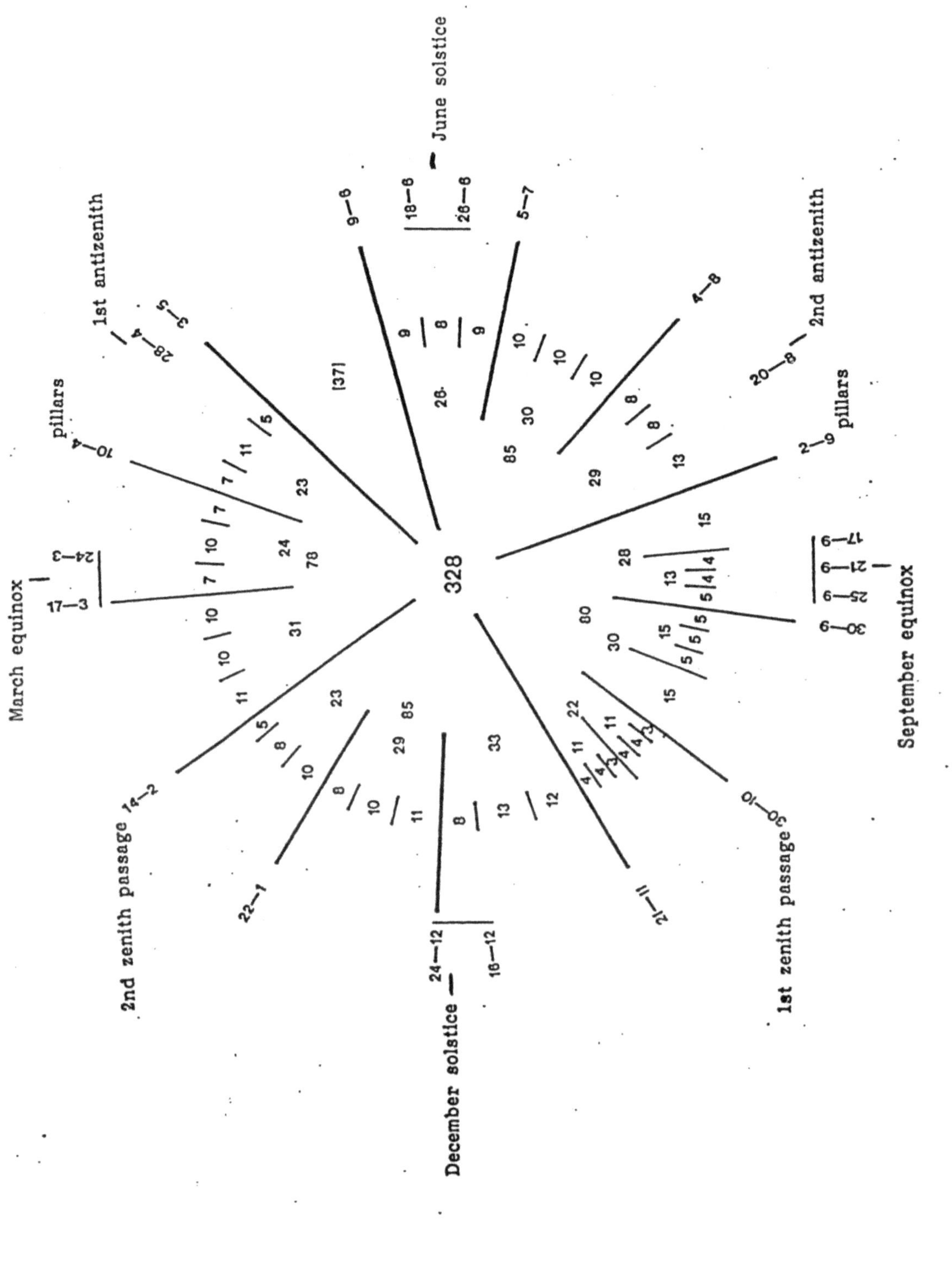

Figure 1 The Ceque Calendar

Urton). Given the prominent position of these two dark cloud constellations in Guaman Poma's list (which are the only identifiable ones and still well-known), it seems again that he is concerned primarily with the importance of the month of Coya raymi in September.

A comparison of the lunar data to those of the Ceque calendar.

We come now to a comparison of the suggested dates for the use of the pillars with those of the calendar as reconstructed from the Ceque system (see figure 1). I refer to my earlier articles for a justification of the latter, arrived at by a completely different argument. The ceque calendar used the period of increase and decrease of visibility of the Pleiades as a means for measuring the length of other astronomical periods. Thus it did not measure a period of 37 days which fell within the period when the Pleiades were invisible, leaving their counting to other methods (Zuidema MS).

In a numerical reconstruction of the ceque calendar (Zuidema in press c) I observed the sequence of periods with the following numbers of days (beginning 9 June): 26, 59 (30 + 29), 58 (15 + 28 + 15), 55 (22 + 33), 52 (29 + 23), 55 (31 + 24), 60 (23 + 37). The fourth and fifth periods, of 55 + 52 days, measure the time going from the first zenith passage (30 October) to the second (14 February). Together with the third (58 days) and sixth periods (55 days) --four periods corresponding to the ones earlier called 1, 2, 3 and 4-- we come to a total of 220 (58 + 55 + 52 + 55) days. The four periods and their total comply, therefore, very well with the characteristic that the Anonymous author mentioned of two of these double-months: the moon of one month always reaching the other. They also support my reconstruction that the same characteristic can be extended to each of the four periods 1, 2, 3 and 4. Behind the rather irregular system of the four actual periods we detect a more regular one of four double-months of 55 days each (4 x 55 = 220). The actual system measured the movement of the sun from first pillar passage to first zenith passage to December solstice--extending period 2 (55 days) 1 1/2 days beyond this date-- to second pillar passage and finally to second pillar passage.

The ceque calendar might also help to find an answer to the question which of the two full moons would be preferred for celebration when both occurred in either of the four periods 1, 2, 3 and 4. In the ceque calendar, one of the two month-like periods in each double-month is always longer than the other. It appears that in period 2 the second moon would be chosen for celebration and in period 3 the first, being the two moons closest to the December solstice. Such a conclusion confirms for period 3 the information we have on lunar celebrations in the month corresponding to January. In period 4 the first moon would have been celebrated, but we lack further information here. 3/

The ceque calendar is of most interest for studying the lunar celebration in period 1. Here it seems that a period of 28 days, either defined as one corresponding to the 28 days of visible moon or as one corresponding to a sidereal-lunar month of 27 1/3 days, is bracketed in between two 15-day periods, each equal to half a synodic month. The 28 day period is subdivided into six periods of 4, 4, 5, 5, 5 and 5 days, which is the closest approximation, in terms of full days, to a regular six-fold division. The two 4-day periods bracket the date of the September equinox. The 28-day period may have corresponded to a movable lunar period of equal length that could begin as early as 2 September and end as late as 30 October, the date of the first zenith passage of the sun.

The ceque calendar may support also another property of the pillar system. In our earlier conclusions concerning the distance between the outer pillars and the distance of all pillars from the Ushnu, Aveni and I had suggested that the sun went through the whole system in about a month. As the two dates of 20 August and 17 September are also marked in the ceque calendar, they may have indicated the earliest and latest possible new moons corresponding to the month of Coya raymi.

But why should the Incas have chosen to bracket the shorter 28-day period within a split,longer, 30(= 15 + 15)-day period and not let the one be followed by the other? I suspect that in the case of this period, where the worship of the moon was of such practical and ritual importance, they may have made another kind of choice than in the other three periods. When the new moon would fall after 2 September, they might have celebrated the next full moon, but when it would fall before 2 September they might have chosen the second one. Thus the full moon to be celebrated for the equinox would always fall around or after that event.

Guaman Poma's text led me to discuss two dates: 29 August and 6 or 7 September. He seems to relate both to the same problem, that of the full moon in relation to the September equinox, even if, in fact, they represent different problems. The date of 29 August is close to that of 2 September and the latter date was chosen in the ceque calendar beginning the two double-months (each shorter than two synodic months) that preceded the December solstice. Guaman Poma may have chosen the day of the death of St John being aware of this problem in the Incaic calendar and possibly of other Andean calendars as well. It may have been an Andean interest to combine the moon-equinox problem with that of the lunar double-months. How the ceque calendar accounts for that combination might be understood as follows.

When Guaman Poma mentions that the sun during the solstices does not move for some days from his seat, he expresses the observable fact that then it does not change its rising and setting points. The ceque calendar defines those periods as of 8 days. But Guaman Poma talks of other seats as well, where the sun rests during other times of the year, and the ceque calendar suggests that the equinoxes were celebrated with similar periods: 8 days around the September equinox and 7 days around the March equinox. Defining the equinoxes as periods and not as days would have enabled the Incas to combine their two lunar interests: those of the moon-equinox connection and of the double-month periods.

They calculated the time of passage through the central pillars as occurring 15 days before 17 September.

Let me come now to the question why the Incas should have been interested in the problem of four shorter double-months. Guaman Poma did suggest the connection of the moon to the stars at night and to the time of the year when the plants were in the ground from planting to harvest. An added reason for such an interest may have been also of a technical nature. We have no information from Cuzco on any lunar celebrations during the time from second pillar-passage, 10 April, to first pillar-passage, 2 September. Even in the case of Inti raymi, where we do know that the temple of the sun was directed towards a sunrise, 25 May, defining the earliest possible new moon related to the June solstice, no information on a lunar celebration is given. The two remaining double-month periods that I did not mention yet, 10 April - 9 June (60 days) and 5 July - 2 September (59 days), are equal to or one day longer than two synodic months. Here, apparently, it was of no concern if or when a lunar celebration was carried out. Elsewhere (Zuidema 1982 b) I suggested that an underlying interest in the ceque calendar was a sidereal-lunar one; one reason being that its total length of 328 days is equal to 12 sidereal months (12 x $27^{1}/3$ = 328). In this way, the ceque calendar extended and completed a sidereal-lunar interest that is more pronounced in the 220 day-period. While this period is close to 8 sidereal months (8 x $27^{1}/3$ = $218^{2}/3$), the other part of the year of $145^{1}/4$ days comes close to 5 synodic months (5 x 29.5 = 147.5). Either 8 sidereal months plus 5 synodic months, or 9 sidereal months plus 4 synodic months are the closest possible approximations to the length of a solar year and the Incaic experts may have been aware of this property of their calendar.

The argument defending an Inca interest in a series of four shorter double-months might clarify also two other issues resulting from this discussion. The ceque calendar, while accounting well for the date (10 april) of the second passage of the sun through the inner pillars does so less for the outer pillars as compared to their first passage. Probably the Incas were not much interested in the moon at this time of harvest anymore, as no chronicler records any lunar feast for that occasion. This fact is the more remarkable as our descriptions for the harvest are no less elaborate than those of sowing and of Capac raymi. According to the ceque calendar, the last lunar feast might have been held before the March equinox; the date after which the sun again followed a lower course than the full moon. 4/ This argument supports my earlier conclusion that the information of the Anonymous on the double-month of March-April should apply to period 4 (14 February - 17 March).

After this comparison between the lunar information presented in the first two sections and the ceque calendar, we can now ask the question if other information in the chronicles becomes relevant about the particular relationship of the moon to the September equinox as reconstructed. I suggested that in period 1 always the full moon after the September equinox was chosen for celebrating that event (which in the particular case of the Inca calendar probably was defined as the period starting 17 September). The chronicler Garcilaso de la Vega (book 7 ch. 6) makes an explicit statement to that effect. While in general I do not trust his information too much when he is not backed up by supporting evidence from earlier chroniclers, we might take him serious here. Our earliest chronicler, Betanzos (ch. 15, p. 73), says that the principal rituals of Coya raymi began with the waxing half moon and Molina (p. 44) gives us to understand that they began four days before full moon. While foreigners had been expulsed from town before or during the preceding new moon, they were reintroduced after those four days. It seems plausible that the reintroduction occurred when the days became longer than the nights and when the sun started to follow a higher course than the full moon. Accepting that the full moon celebrations took place after the equinox, it would make a statement of the chronicler Murua (p. 108) less puzzling. He claims that they did not occur in September but on 1 October. He refers to a date close to the full moon after the equinox.

I can conclude this section defining the calendrical significance of the pillars. It seems most plausible that the Anonymous author was right when he stresses, first of all, their importance for timing in relation to planting. Early planting, at altitudes above the Cuzco valley, was started when the sun passed the first outer pillar, and general planting when it set in between the central ones. These dates were probably set out of practical necessity. However, the Incas had to justify these dates within the context of a well argued calendar and against the background of solar observations that had astronomical and cosmological significance; this they would have had to do in order to communicate about the pillar observations to people elsewhere. They used a lunar theory predicting, from a new moon, the first full moon after the September equinox. We have no field data from Cuzco on a solar observation of the equinoxes supported by ethnohistorical information. I suspect, however, that the outer pillars did not serve the only purpose of indicating a month-long period around 2 September; they also helped to define the dates of the antizenith sunsets and of the equinoxes. The date of 20 August as indicated by the ceque calendar is only two days after the antizenith sunset and its date of 17 September only four days before the equinox.

Coya raymi and Guaman Poma's description of a personal ritual.

By combining the issue of the date of 29 August with those of Coya raymi and the moon near the September equinox, Guaman Poma allowed us to consider the problem of the pillars in Cuzco from a more general perspective. My conclusions about the dates of the sun's passage through the pillars seemed to be confirmed by the calendar as derived from the ceque system. This system was in the first place one of ritual space. Thus we can confirm, on the one hand, that the ceques related to the South were taken care of ritually by groups of non-Incaic origin, and on the other, that the months of September, October and November derived their names from non-Inca groups. September, Coya raymi, was named after the queen in honor of women in general; October, Uma raymi, after one pre-Inca group of the valley of Cuzco; and November, Ayarmaca raymi, after another. As the ceque calendar and its ceques to the South supported best my hypothesis about the importance of the period from first pillar passage (2 September) to zenith passage (30 October), we can ask the question if there was a special significance to the relationship between the South and these months. If so, it would constitute a major support for the calendrical interpretation of the ceque system as offered.

The last part of Guaman Poma's text seems to lead to an answer of this question. We read about an event that had happened to a woman and her deformed child, told to him by an informant of the village of Asque, through which he once had passed. It was a non-calendrical event, but used as an exemplary story illustrating the significance of the month of Coya raymi and its rituals in Cuzco. The story places these in a wider Andean context and presents us in the end with an argument why in the ceque calendar the southern ceques accounted for the rituals of September and October. Because of its crucial place in Guaman Poma's treatise on astronomy and the calendar, let me first translate the text in full:

> The use of fasting and the vigil and the ceremony that the Yngas keep till today in the village of Asque:
> The indians carried out ceremonies for lightning whom they call Santiago (Saint James). Before they called him Yllapa, and with another name curi caccha (golden clap of Thunder). It is because in that village a child was born with a sunken nose, one that they call cacya cinca ("nose (obtained because) of the Thunder". And they shut the woman and her son up (in a dark room), calling him "son of Santiago", where nobody spoke to her except an old man. And they made her fast from salt, agi (spanish pepper) and meat; they only gave her white maize to eat. And she did not see the sun, nor the moon.
> After this month, all (the people of) the village came together and they made some ropes twisted towards the left and of which they made the ends burn. They all whipped the woman with these burning ropes and threw her out of the village. And (with her face) covered she fled from the village with her son to a small mountain. And when they whipped her they said: "Get out of this village, wife and son of Santiago yllapa"!
> And she went to the small mountain and they say that she was there for a month and her son died and they buried the child. On top (of the grave) they buried alive a small black llama and other dirty things. There they sacrificed to the Lightning, Santiago. And they brought the woman back to the village, richly dressed and with their taquies ("dances"). And they were drunk for five days till they fell down, the whole village, and when the (Catholic) priest was absent. And the "cantor" (singer) was there and it was his student who told me (the whole story).

In another context, that of "common sorcerers" (p. 276(278)),

Guaman Poma had already talked about such children with a sunken nose:

> Other priest-sorcerers order (what to do) when two children are born from one womb or with sunken noses guaca cinca ("huaca nose"), or who are born with the feet first yayuma, uizama, aualla, of whom they say today that they are son (children) of Santiago, son of yllapa and of curi. They do not want to visit them and (they make them) fast from salt and other food and (during that time) men do not sleep with their wives.

Children with sunken noses are placed in the same category as twins and the rituals for the first are the same as those carried out for the second (Arriaga, Avila, Guaman Poma). The parents were first shut up in a dark room and then chased out of the village to a high place in the middle of night. Guaman Poma describes the burning ropes in a similar ritual act as carried out during the night of the first new moon in the month of Coya raymi. People did it as a jocular act for cleaning each other from the evil and illnesses in their bodies. The ritual was preceded by expelling from town all foreigners and people with physical defects, and it was followed by that of 400 warriors running out of town into the four principal directions. The two main rivers, Villcanota and Apurimac, received the evil and the illnesses that had been thrown out of town.

On top of the burial of the child a small black llama was placed alive. This act reminds us of the custom carried out in Cuzco during the month corresponding to October (Guaman Poma p. 254-5(256-7)). A black llama was tied to a pole in the plaza, starved from food and drink, in order to make it weep for rain. In general, black animals were killed, weakening the enemies that they were thought to represent. But in the October ritual the animal was not killed. As Guaman Poma explicitly mentions that the small black llama on the boy's grave was alive, his claim seems to be that here too the animal was not killed, making the two rituals even more comparable.

Giving birth to a child with a sunken nose is a non-calendrical event. But it is obvious that the burial of the child became integrated into a calendrical context. Still today, fetuses and unbaptised children are buried at the foot of the crosses that are erected on mountain tops of the horizon around a vilage and its agricultural lands (Harris 1982; own field observations from Ayacucho). These crosses are brought down during the Feast of the Cross, 3 May, adorned with harvest products. They are returned in relation to the next agricultural season, but this may happen either within a week, or in September when planting starts, or in November when the heavy rains begin (Urton 1986). The crosses protect the crops, but the ritual described by Guaman Poma also suggests that different transformations occurred at the place of the burial. The status of the mother was transformed from an objectionable to an honored one. While in the ritual of the burning rope her head was covered --reminding us that in the temple of the Thunder in Cuzco people could not see the face of this god's statue--, and while in the month before she was kept in a dark room without seeing the sun or the moon, she was received with much festivities when she came back from the mountain. In August rituals were carried out, fearing the possible nephastous influence of the Thundergod in its various athmospherical manifestations (Cobo p. 216), but in October the benevolent rains were attracted as another manifestation of the same god. Today it are the crosses that operate this transformation and we may conclude that burials like of this "child of the Thunder" had a similar function before.

Guaman Poma divides his ritual over three months and I have compared the three parts to the calendrical rituals of August, September and October. Both sequences do not overlap exactly. In his example the black llama-ritual occurs first and the one honoring the woman after, while in the calendrical one September is the month honoring the Moon, the Queen and all women and October the month of the black llama-ritual. Nevertheless, we can compare his ritual of the burning rope to that of the burning torches made of maize leaves and used during the new moon around 2 September. His ritual of bringing back the woman coincides with the invitation back into town of the foreigners, these being honored during the first full moon after the September equinox.

Molina, in his description of Coya raymi, had said that it were the illnesses caused by the first rains that were thrown out of town. The argument is not convincing as by then the rains had not arrived yet. A similar ritual was held in Central Peru near the June solstice, first, as a protection of the potatoes that then were harvested and, second, to exorcise the people coming back to the village after having served the Spaniards and eaten foreign food (Burga p. 187). The second argument mentioned here would have been valid also when outsiders returned to Cuzco after the full moon of Coya raymi, an event that Guaman Poma relates to the September equinox, when a lunar eclipse as illness of the full moon was mostly feared.

Guaman Poma's description widens our perspective of the calendrical feast also in another way, now relating it to space. The event caused by the birth of the child with the deformed nose, was only a particular example of burials of twins, of children born with the feet first, etc.. Hernandez Principe explains the general importance of such burials in the case of the village of <u>Recuay</u>, Central Peru (Duviols 1986 p. 487; Zuidema 1973 a, b). As part of his description of its social organisation he says:

> They worshipped the house of the bezoar stones which they call <u>Illahuasi</u> offered to the Lightning for the increase of the llamas and they had for that a deposit. Each head of a family, whom they call <u>churi</u> ("son") kept its own, well-known location above the old (preconquest) village near to the place of worship of the Lightning where they had their household gods that they call <u>conopas</u>. There are their sacrifices of llamas in circular structures where they hide the ashes and the fat of llamas.

> In lines like of the huairas ("smelting ovens") of Potossi, each (family) kept their fetuses and chuchos and chacpas born with the feet first and twins ("those born from one womb") offered to the Lightning or Sun or Venus ("lucero") according to divination of the sorcerers. Next to them they had the sacrifices of the llamas by which they consulted them (the twins etc.).

Hernandez Principe mentions the same kind of burials, like of twins and children born with the feet first, as Guaman Poma, who adds to these the children with sunken noses. They both refer to a specific mountain outside the village, but Hernandez Principe combines this element of the ritual with another, similar to the ritual use of the ceque system in Cuzco, where evil and illnesses (like the child with the sunken nose) were thrown out along the directions of the ceques. The burials of twins etc. were done according to directions, each belonging to a specific family, emanating from the top of the mountain.

Elsewhere I have argued that the mountain used in Cuzco for similar purposes was that of Anahuarque, some 37° East of South on the horizon of the temple of the Sun, and that its direction served to observe the rise of the constellations of the Southern Cross and of Alpha and Beta Centauri, known as the "Eyes of the Llama" (Zuidema and Urton). This direction, together with that of the setting of these constellations was used in the Andes (like in other cultures of the southern hemisphere) to define the South direction. While these quasi-circumpolar constellations were worshipped especially during the months around the September equinox when they had their lower culminations around midnight, the descriptions of Guaman Poma and Hernandez Principe provide us with an argument relating in Cuzco the South to the September equinox.

Notwithstanding the fact that twins etc. must be considered as offspring of the families that buried them, they were worshipped as a category equal to that of ancestors. Albornoz (p. 196) classifies twins and other uncommon births as huacas, together with mummies of ancestors, and calls them both illapa; a word that in its first meaning refers to lightning. We conclude that, while in Guaman Poma's story the child was worshipped as an illapa or ancestor, his mother, returning from his grave, was celebrated because of the fertility that women contribute to the village. In the case of Cuzco, it were primarily the women of conquered subjects who were considered in that role (Murua p. 107-9; Bertonio, under the word pacoma; Zuidema 1986 p. 74). With this background information it will be possible to understand the Inca origin myth in its particular meaning for the rituals of the month of Coya raymi in September.

When Sarmiento de Gamboa, following Betanzos, transforms the theme of the defense of Cuzco by king Pachacuti Inca into the causal element of the foundation of Cuzco as a city, he describes (ch 12, 13) this act as an elaborate ritual involving the ten constituent panacas as its political components, the twelve months, the six sacred mountains including Mt Anahuarque, and the four major feasts including those of the solstices, of planting and of harvest. But his first acts are the building of terraces for sowing, the placement of the ushnu and the pillars for defining the times of planting and harvest, and a visit to the cave of Tambotoco "the window or cave Tambo" near the town of Pacaritambo "the place of origin". His construction of the ancestral system --including 7 mummies and the statues of 3 ancestral deities-- was represented as a re-construction and it was meant as a consecration of his claims to sacred kingship, being reborn in the same cave from where the first ancestors had emerged. We can read the story of Guaman Poma as a similar reenactment. The same close relationship that he establishes between his personal story and the feast of Coya raymi is also found in Sarmiento's account where he combines the acts of building the terraces in Cuzco and of the system of pillars, used for the timing of planting and harvest, with the king's entrance into the cave of origin.

Pachacuti Inca's entrance into the cave can be seen also as a reference to the origin myth in a more particular way. One of the brothers coming out of the cave, Ayar Cachi, was sent back by the others, fearing his strength, in order to fetch a golden beaker, called tupa cusi, and the seeds to be sown in the valley of Cuzco. Ayar Cachi was, however, emprisoned in the cave by a servant and from there he caused an earthquake. Sarmiento also mentions another version of this episode. Here, Ayar Cachi was buried by his brother Manco Capac in the valley of Cuzco itself, while the latter wept as an act of attracting the fertility to the valley; weeping like people did in the month of October. Ayar Cachi was an underworld giant (huari) and, as an ancestor who was sacrificed and buried as seed, he was intimately related to the fertility of the earth in Cuzco. Pachacuti Inca, by visiting the cave, identified himself with Ayar Cachi. His wife, the queen, took her name Mama Anahuarque from the mountain of the same name and it were his more than hundred children in women obtained in war that brought their fertility to the valley (Zuidema 1986). 5/

The theme of Tambotoco is at the core of the Incaic interests in September as the month related to the fertility of the earth. The cave is found at a distance of ± 30 kms straight South of Cuzco; a direction obtained by way of observing the rising and setting points of the Southern Cross and the "Eyes of the Llama". It remains, of course, a question to decide if Tambotoco was chosen as the cave of origin because it was located South of Cuzco or if this direction was accidental to its function. I am inclined to opt for the first possibility and to refer to the South as the cosmological direction of origin. It conforms to a concept still current in the nearby town of Chinchero, also of Incaic origin, where the South is called Ñawpanchis, "our past" (C. R. Franquemont p. 32). 6/ The South in Cuzco was not only the direction of the ancestors and from where fertility came, but also the one to where the dead returned, to the mountain of Coropuna much further away. This idea is also still alive today. The Incas had good calendrical and astronomical reasons, grounded in the local culture of their capital, to relate the southern ceques to the time around the September equinox.

Conclusions.

I built my argument about the calendrical significance of the pillars up around three types of information that we can accept as of prespanish origin and that give us a possibility to check out empirically their validity. They concern the dates of the solstices and of the passages of the sun through zenith; the lunar celebrations related to months that the Spaniards identified as September and January; and the information on the fixed double-months having less than 59 days. First I concluded that the dates of the December solstice and of the zenith passages were meant to separate calendrical periods whose lengths were also expressed in terms of Incaic months. From there I drew the conclusion that the periods between the zenith passages and the pillar passages were of the same nature and length as the first two ones. Thus I came to the dates of ± 2 September and ± 10 April for sunset in between the central pillars. Next, I compared these results with those reached on the basis of the ceque calendar and of the calendrical and astronomical treatise by Guaman Poma. In both comparisons I could come to completely satisfactory results. An added value of the second comparison was that the information of Guaman Poma did not derive from Cuzco, although he himself compared it to that on the Inca calendar. 7/ The reason for erecting the pillars was not just a local one of Cuzco but was based on more general calendrical principles; the calendar integrated the local information on the zenith passages and the antizenith sunsets but was not dependent on it.

Comparing the different sources of information we begin to understand more about an Andean theory of the moon in relation to the solar year. On the one hand, the Incas were interested in a full moon around the June solstice, in a full moon after the September equinox and in a full moon after the December solstice; the Cuzco data do not allow us to say anything about a full moon after the March equinox, and I suspect that they only had a minor interest in the problem. The difference in solar observation during the solstices and the equinoxes led to different techniques in establishing the lunar connection. As the sun returns on its path during the solstices, only one observation has to be made for defining two dates before and after that event, respectively; from there the moon around the solstice could be calculated. But the sun around the equinoxes continues its path and here a system of three pillars was necessary.

On the other hand, the Incas were also interested in separating the full moons to be celebrated throughout the year from the ones in which they had less interest. For this purpose they had a system of double-months in which four were shorter than a double synodic-month and two equal to or longer. The ceque calendar suggests, however, that even in the case of the shorter double-months one month was equal to or longer than a synodic month. Thus there were six longer months spread out over the year with a total lenght equal to half a year, and six shorter months that, together with the extra 37 days not accounted for in the ceque calendar, corresponded to the other half.

One early chronicler, Polo de Ondegardo, mentioned that the Incas had a system of synodic months, to which months could be added one or two days "according to the phases of the moon". He heard something about the Inca calendar, but did not understand its technicalities. From his wording we conclude, at least, that these months were, in fact, not synodic at all. I disagree, therefore, completely with the reconstruction of the Inca calendar as offered by Ziolkowski and Sadowski (1982-4). Although they are aware of Polo's observation, they disregard its consequences. They also do not take into account, first, the observation of the Anonymous about certain fixed double-months being shorter than two synodic months and, second, the calendrical reasons why the Incas made their various observations of the sun throughout the year. They reconstruct a synodic lunar calendar beginning with a full moon around the June solstice, including an extra 13th month about every third year. If such a construction were true, there was no reason for erecting the pillars and we would have to accept that the dates when the sun went through zenith had no calendrical function. There is no evidence for their Deus Ex Machina of an extra month every three years. Their reconstruction, though suggesting a location in the year for the Quechua month names as used in colonial times, has little to do with the system of the prespanish calendar and its problems.

Notes.

1 Mt Picchu as a huaca in the ceque system (the seventh huaca on the eighth ceque of Chinchaysuyu) was also known as Sucanca, a word referring to a point on or near the horizon used for observing a sunrise or -set. The mountain is said to be named so because of the central pillars. But the only two other places called Sucanca were near the far eastern horizon, where the DSSR and the zenith sunrise were observed without the use of pillars. I suspect, therefore, that Mt Picchu was called Sucanca only for observing the antizenith sunset and not for the sunset in between the central pillars. Because of this ambiguity in the use of the term Sucanca I will refrain from using it in this article.

2 Further support for this argument is given in section 3.

3 It may be of interest to observe that the Anonymous author, when giving his information on the two double-months, in each case assigns a name -cuzqui, "to plow", to the month that in my comparison to the ceque calendar would be the shorter one of the two. No other chronicler follows such a system, although it is true that Chaupi cuzqui "the middle time of plowing" corresponds to the time in November, called cuzqui mitta (Gonzalez Holguin) when hoeing was carried out, and that Haocay cuzqui reminds us of the fact that plowing of new lands used for potato cultivation was done in about February - March (Guaman Poma p. 241(243)). All chroniclers agree that the last month before the one of the June solstice was called Hatun cuzqui "the large or the long month of plowing", done after harvest. The Anonymous says of Haocay cuzqui, comparing it to April, that this month was taken together with March "because the occupations in these two months continued till the end of both months". It may be that all the months -cuzqui had an additional character: Hatun cuzqui as the period of 37 days in relation to the period of 328 days in the ceque calendar, and the two other months in relation to the other, longer months in each of their double-month periods.

4 The outer pillars probably were of no use any more around 10 April. I will deal with this problem and that of the months after the March equinox in a separate article.

5 Cummins (1988 p. 209-215, note 51), observing this connection between Ayar Cachi, Tambotoco and fertility, chooses as a reading of the name of the golden beaker, tupa curi "royal golden ...", based on a faulty transcription in one case (Santacruz Pachacuti p. 214; cf. the manuscript) for a name that otherwise always is written as tupa cusi. The latter word means "golden luck and contentedness", luck being considered "in terms of works and temporal things" (Gonzalez Holguin: cusi; see also under atau). This meaning is more in line with the state of mind that people were expected to be in during the month of Coya raymi.

6 The four cardinal directions here are called: East, Inti lluqsimuna "sun coming out this way"; West, Inti haykupuna (haykupusan) "sun going away"; North, Q'ipanchis "that which

follows us"; and South, _Ñawpanchis_ "our past".

7 At the beginning and end of the first part of this treatise he says that he got his information from a calendrical expert, 150 years old, in the village of _Uchucmarca_ of the province of _Lucanas_. The story in the second part came from another village, but probably one that was not too far away. There exists a modern village with a similar name, _Asqui pata_, in the province of Soras and in the same department of Ayacucho as Uchucmarca.

References

Albornoz, Cristobal de

see: Duviols, Pierre (1984)

Anonymous

"Discurso de la sucesión i gobierno de los Yngas." In _Juício de límites entre el Perú y Bolivia_, _prueba peruana_, ed. V. Maúrtua, vol. 8 Chunchos: 149-165. Madrid. (1906)

Attwater, Donald

The Penguin Dictionary of Saints. Hammondsworth: Penguin. (1983)

Aveni, Anthony F.

Skywatchers of Ancient Mexico. Austin: University of Texas Press. (1980)

"Horizon Astronomy in Incaic Cuzco." In _Archaeoastronomy in the Americas_, ed. Ray A. Williamson. Los Altos: Ballena Press. (1981)

Bertonio, Ludovico

Vocabulario de la lengua Aymara (1612). La Paz: Don Bosco. (1956)

Betanzos, Juan de

Suma y Narración de los Incas (1551). Madrid: Atlas. (1987)

Burga, Manuel

Nacimiento de una Utopía. Muerte y Resurrección de los Incas. Lima: Instituto de Apoyo Agrario. (1988)

Cobo, Bernabé

Historia del Nuevo Mundo (1653). B.A.E. vols 91-2. Madrid: Atlas. (1956)

Cummins, Th. B. F.

Abstraction to Narration: Kero Imagery of Peru and the Colonial Alteration of Native Identity. Ph. D. Thesis, University of California, Los Angeles. (1988)

Duviols, Pierre

"Albornoz y el Espacio Ritual Andino Prehispánico. Instrucción para descubrir todas las guacas del Piru y sus camayos y haziendas." Revista Andina 1, 169-222. Cusco. (1984)

Cultura Andina y Represión. Procesos y Visitas de Idolatrías y Hechicerías Cajatambo, Siglo XVII. Cusco: Centro "Bartolomé de las Casas". (1986)

Garcilaso de la Vega (El Inca)

Comentarios Reales (1609). Buenos Aires: Emecé. (1945)

Gonzalez Holgúin, Diego

Vocabulario de la lengua general de todo el Peru llamada lengua Qquichua o del Inca (1608). Lima: Imprenta Santa María. (1952)

Guaman Poma de Ayala, Felipe

Nueva Crónica y Buen Gobierno (1615). Madrid: Historia 16. (1986)

Harris, Olivia

"The Dead and the Devils among the Bolivian Laymi." in Death and the Regeneration of Life, eds. Bloch, M. and J. Parry. Cambridge: Cambridge University Press. (1982)

Molina, Cristobal de

Relación de las fábulas y ritos de los Incas (1573). Lima: Librería e Imprenta D. Miranda. (1943)

Murúa, Martin de

Historia del Orígen y Genealogía Real de los Reyes Incas del Perú. Madrid: Instituto Santo Toribio de Mogrovejo. (1946)

Santacruz Pachacuti Yamqui Salcamaygua, Joan de

Relación de Antiguedades de este Reyno del Perú (1613). Reproduction of edition M. de la Espada. Tres relaciones de antiguedades peruanas, pp. 207-281. Asunción del Paraguay: Guaranía. (1950)

Sarmiento de Gamboa, Pedro

Historia de los Incas (1572). Buenos Aires: Emecé. (1947)

Urton, Gary

At the Crossroads of the Earth and the Sky: an Andean Cosmology. Austin: University of Texas Press. (1981)

"Calendrical Cycles and their Projections in Pacariqtambo, Peru." *Journal of Latin American Lore*, 12, no. 1, 45-64. (1986)

Voragine, Jacques de

La Légende Dorée. Paris: Garnier-Flammarion. (1967)

Ziolkowski, Mariusz S. and Robert M. Sadowski

"Los problemas de la reconstrucción de los calendarios prehispánicos andinos." *Estudios Latinoamericanos* 9. (1982-4)

Zuidema, R. Tom

"La parenté et le culte des ancetres dans trois communautés péruviennes: un compte-rendu de 1622 par Hernandez Principe." in *Signes et Langages des Amériques, Recherches Amérindiennes au Quebec*, vol III, nos. 1-2, pp. 129-145. Montréal. (1973 a)

also published as:
"Kinship and Ancestorcult in Three Peruvian Communities; Hernandez Principe's Account in 1622." *Bulletin Institut Francais des Etudes Andines*, vol. 2, pp. 16-33. Lima. (1973 b)

"The Inca Calendar." in *Native American Astronomy*, ed. A. F. Aveni, pp. 219-259. Austin: University of Texas Press. (1977)

"El Ushnu." *Revista de la Universidad Complutense* (Madrid), vol 2, no. 117, pp. 317-362. (1980)

"Inca Observations of the Solar and Lunar Passages through Zenith and Anti-zenith at Cuzco. In *Archaeoastronomy in the Americas*, ed. Ray a. Williamson. Los Altos: Ballena Press. (1981)

"Catachillay: the role of the Pleiades and of the Southern Cross and Alpha and Beta Centauri in the calendar of the Incas." in *Ethnoastronomy and Archaeoastronomy in the American Tropics*, eds. A. F. Aveni and Gary Urton. Annals of the New York Academy of Sciences, 385. New York, pp. 203-29. (1982 a)

"The Sidereal Lunar Calendar of the Incas." in *New World Archaeoastronomy*, Ed. A. F. Aveni. Cambridge: Cambridge University Press, pp. 59-107. (1982 b)

La Civilisation Inca au Cusco. Paris: Presses Universitaires de Frances / College de France. (1986)

"The Moieties of Cuzco." in The Attraction of Opposites, eds. Maybury Lewis, D. and U. Almagor. Ann Arbor: University of Michigan Press. In Press a.

"At the King's Table." Anthropology and History, volume dedicated to the theme of The King, ed. J. C. Galey. Paris. In Press b.

"A quipu Calendar from Ica Peru; with a comparison to the Ceque calendar from Cuzco." in Acts of the second Oxford conference on Archaeoastronomy. ed. A. F. Aveni. Cambridge: Cambridge University Press. In Press c.

"Las Pleyades y la Organización Política Andina." Paper presented at the meetings of the Asociación Peruana de Etnohistoria. Lima, 1981. MS

Zuidema, R. Tom and Gary Urton

"La Constelación de la Llama en los Andes Peruanos." Allpanchis Phuturinqa, 9, 59-119. (1976)

DIRECTIONALITY AS A CONCEPTUAL MODEL FOR ZUNI EXPRESSIVE BEHAVIOR

M. JANE YOUNG

Perceptions of the motions of the sun, moon, and stars, as well as concepts of time and space form an organizing principle for much of the expressive behavior of the Zuni Indians of western New Mexico. They regard the order apparent in the cosmos as a paradigm -- a model for conceptually ordering their verbal and visual art. They do not, however, consider all cosmological phenomena to be equally important in this respect; as is common with agriculturalists, the Zunis devote most of their attention to the apparent recurrent motion of the sun. The pre-eminent position of "Our Father the Sun" in Zuni thought is illustrated by the fact that the Zuni word for "life," tek'ohannanne, is the same as the word for "daylight."

The Zunis appoint an officer whose main function is to watch the sun's daily points of rising and setting from a special sun-watching station and to announce the times for events -- he is, therefore, the keeper of the calendar.[1] This does not imply, however, that the other members of the pueblo pay no attention to the sun. Incidents are known in which the sun watcher erred and the other townspeople ably corrected him (Parsons 1917: 296-299). Members of the community pray to the Sun Father each morning as he rises in the east. They believe that when the sun rises he inspects the character of the prayers and offerings addressed to him and takes them with him when he goes in the evening to his home in the west.

Although the Zunis note the sun's daily motion -- its rise and set positions along the horizon -- with keen interest, it is the yearly motion of the sun that forms their basic conceptual model. Zunis regard the extreme "swings" of this annual cycle -- the winter and summer solstice rise and set positions -- as the most critical times of the year, for it is then that the sun appears to stand still for four days during his journey along the horizon. To encourage "Our Father the Sun" to move again, particularly to assure that he turns around and continues his journey back along the horizon, the Zunis perform complex and vitally important rituals. The most elaborate of these occurs at the winter solstice when the sun is at its southward pause, perhaps because if the sun did not move northward again the fields would stay frozen and unsuitable for planting. The ceremony occurring at this time marks the beginning of the calendrical year and the cycle of ritual performances for the Zunis. The summer solstice ceremony is an abbreviated version of the winter solstice ceremony.

The apparent annual motion of the sun along the horizon serves to divide the Zuni year into two parts, each lasting for six months and each beginning when the sun reaches the solstice position and turns around. The winter and summer solstice ceremonies that occur at this time are both called 'itiwana, "the center," although it is the winter solstice that marks the starting point of the new year cycle. In a kind of symbolic economy, the Zunis repeat the names of the months so that the first month of the winter season has the same name as the first month of the summer season, and so on, counting only six names in all. Winter and summer are thus more than complementary; they are reflections of one another. Nevertheless, distinctive ceremonies and activities characterize each half of the year. There are, for instance, a winter cycle and a summer cycle of ritual dances; certain narratives may be related in the winter but not the summer, and so on. These examples illustrate that the idea of duality or balance is important in Zuni thought, a duality that is expressed in dichotomous terms such as: morning/evening or sunrise/sunset, summer/winter, above/below. The Zunis consider the Sun Father to have two houses that he visits in the course of his daily journey: one in the eastern ocean and one in the western ocean. They associate this morning/evening duality with two deities who play a central role in the Zuni origin myth: the Twin War Gods. They are sons of the Sun, the Morning and Evening Stars, one going before and one following the Sun. Taking pity on the people and desiring their prayer offerings, the Sun asked these Twins to lead the people from the four underworlds to the surface of this earth back in the time of "the beginning." The principle of duality is significant not only in and of itself, however, but also because it gives rise to the Zunis' quadripartite view of the world, achieved by combining the solar positions -- yearly and daily motion, winter and summer solstice sunrise and sunset -- so that the entire annual cycle is represented (see Figure 1).

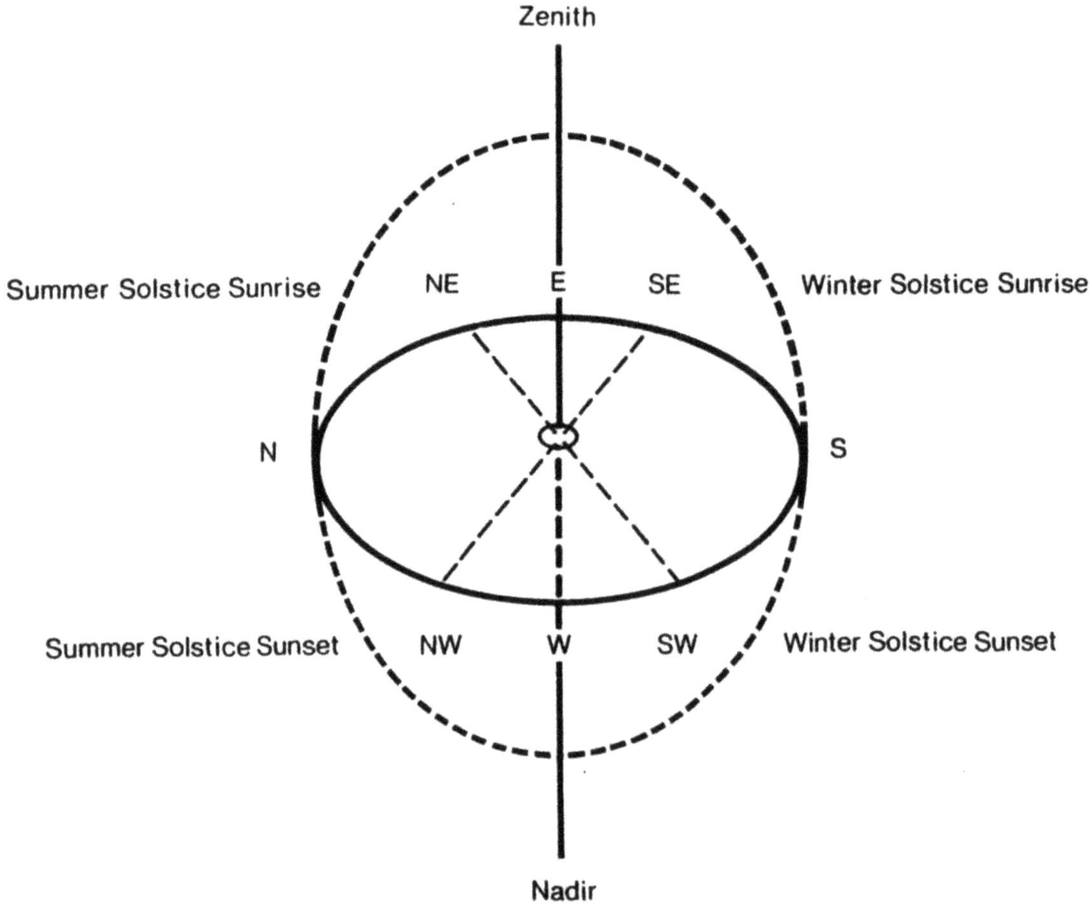

Figure 1. Schematic drawing of the circle of the earth's horizon with solstice rise and set points marked (the inner circle and dotted lines could represent the heart and legs of the Water Skate described in part of the Zuni origin myth). The inclusion of the zenith and nadir extends the circle into a sphere. Drawing by Murray Callahan.

The number four, associated with the solstices as well as sunrise and sunset, and hence directionality, is both sacred and ubiquitous in Zuni mythology and cultural symbolism (Benedict 1935: 1-276; Bunzel 1932: 547-609; Cushing 1896: 321-447; Hymes 1980: 7-51; Parsons 1923: 135-62; Stevenson 1904: 20-61; D. Tedlock 1972: 225-298).[2] According to the Zuni origin myth, the people traveled through four underworlds to reach the surface of the present world where they were, at first, blinded by the light of the sun. Then they spent four years (or, in some versions, four days) searching for "the Center." It is significant that the Zuni word for this surface of the "earth" or "Earth Mother" (that is, the ground in which seeds are planted), 'awitelin tsitta, derives from the word for "four," 'a:witen, for it is here that the four directions begin to take on importance. Directionality, or location in time and space, was not a feature of life in the sunless underworlds. Furthermore, as noted above, Zunis perceive the sun as standing still at the solstices for four days rather than three or five. The number four is also central to Zuni ideas about reincarnation: they believe that, with the exception of young children and certain priests, each Zuni person is born four times and dies four times. After the fourth death Zunis either return to the hole where they emerged from the four underworlds to this earth in the time of "the beginning," or find that they have descended one of those underworlds with each death, arriving with the fourth and last death at the lowest of the four underworlds where the Zunis originated. At this point they may become a member of one of the species of animals (raw beings) they had an affinity for in life and return to the world of the living (D. Tedlock 1979: 508). Finally, the number four serves as an organizing principle for most other aspects of Zuni life. According to Ruth Bunzel, four is the major literary and ceremonial model (important events are repeated four times; ritual actions are often carried out in a four-part sequence) as well as an ideal pattern in pottery : when discussing their painting, Zuni potters "express a preference for the number four," but it is an ideal that is not always realized (Bunzel 1972: 23).

In Zuni ritual practice, too, this number is significant. For instance, when praying to the sun, Zunis scatter sacred cornmeal in the direction of the four solstice points. These are the four points on the horizon approximating the semicardinal directions of northeast, northwest, southwest, and southeast (their usual ritual order) where the winter and summer solstice suns rise and set and appear to stand still (Harrington n.d.).[3] Located at these four points are four oceans that encircle the earth so that it becomes their " Center"; each ocean contains a mountain that is the color of that particular direction. The importance of these directions is established in the origin myth: the "Center Place" was finally found when the Water Skate, given magical powers of extension by the Sun Father, stretched out his four legs to those same four solstice points (see Figure 2). The place where his heart then rested marked the Center, the heart of the Earth Mother. This is the place where the present-day village of Zuni was built (Bunzel 1932: 601-602; Cushing 1896: 428-429; Stevenson 1904: 46; D. Tedlock 1972: 278-280). Finally, at the center of the village itself, directly under the spot where the heart of the Water Skate rested, lies yet another center. On a permanent altar in the fourth underground room of the house of the chief priesthood is a heart-shaped rock, "the heart of the world," which has within it arteries reaching towards the four solstice positions, just as did the legs of the Water Skate (Bunzel 1932: 514; Stevenson 1904: 46). One can conceive of these various centers, then, as a series of concentric circles, all surrounding this ultimate center, "the heart of the

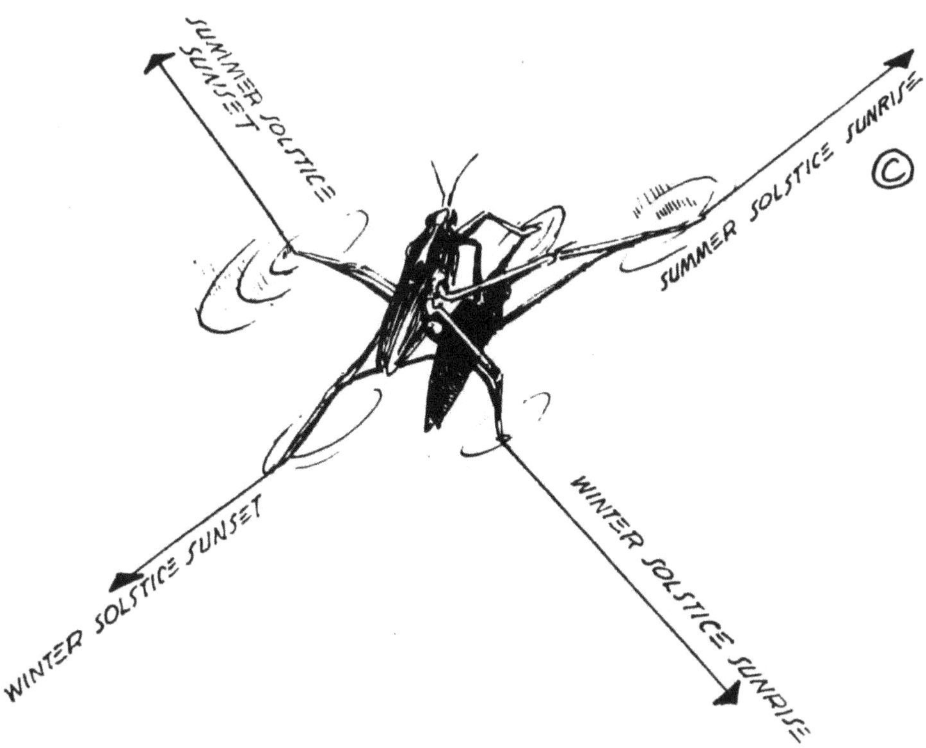

Figure 2. Water Skate finding the center. Drawing by Snowden Hodges.

world." In addition, the Zuni view the universe as consisting of a definite number of vertical layers: there are four underworlds, oriented towards the nadir, and each associated with a tree; then comes the surface of the earth (the world currently occupied); finally, there are four upperworlds, oriented towards the zenith, and each associated with a different kind of bird. Although totalling nine levels, Zunis perceive them as four on either side of the central layer: the familiar world, the beginning point in the cycle of reincarnation (D. Tedlock 1979: 499).

With the addition of these various "centers" to the sacred number four, we now have the number five -- four plus the center. One may say, in fact, that the number five is derived from the two major cosmological principles of the Zuni: 'itiwana, "the center," and 'a:witen, "four." The directional scheme of four plus the center can be represented visually by the same diagram that illustrates the sun's apparent annual motion (Figure 1). Stretched with limbs extending to the solstice points, the Water Skate appears to form a figure X , with his heart at the center (Figure 2).

As with the number four, the number six is ever-present in Zuni ritual. Its origins lie in the Zunis' recognition of the three-dimensional nature of the world in which they live. Thus, the zenith and nadir are combined with the four solstice points located on the horizon that form the seemingly planar surface of the earth. For instance, the Zunis associate each of the six sacred kivas with one of these directions. They link the six Shalako kachinas (major figures in the "coming of the gods" ceremony that occurs prior to the winter solstice) and six pairs of Salimobia ("warrior" kachinas, two of whom accompany the Shalako kachinas) to those kivas. When these kachinas come to the pueblo at the time of the Shalako ceremony, they dedicate the six houses (called Shalako houses) that have been built or remodeled in the past year. Furthermore, the Zunis relate the directions to life-bringing water, as well as to the sun: this cosmological scheme includes six rain priests of the six directions "who take the form of clouds, rainstorms, fog and dew when they leave their homes" (D. Tedlock 1979: 400), six rain-bringing winds, six bow priests who make lightning and thunder, and six different varieties of "water bringing birds" (Stevenson 1904: 22, 1915: 89; D. Tedlock 1979: 499-500). Finally, associated with each direction and symbolizing it is a color.

Animals, too, are drawn into this network of symbolic relations. According to Zuni mythology, six Beast Gods guard the world. As illustrated in the table below, each of these animals is positioned at one of the four solstice points as well as the above and the below, and is colored accordingly.

NE	summer solstice sunrise	Mountain lion	yellow
NW	summer solstice sunset	Bear	blue
SW	winter solstice sunset	Badger	red
SE	winter solstice sunrise	Wolf	white
above		Eagle	many colors
below		Mole	black

TABLE 1: Ritual Associations at Zuni -- usual order.

Sometimes Knife-Wing (the mythic being with wings and tail of knives) rather than the eagle is associated with the zenith (Bunzel 1932: 784; D. Tedlock 1979: 500).[4] Figure 3 depicts the Beast Gods as painted in a wall mural of one of the Zuni religious societies.

Although the number six is significant in itself, there is an implied center in all of these examples; with the addition of this center, the ritual number becomes seven. This can be seen in Figure 1 if one imagines that the circle of the horizon is a slice of a sphere, the axis of which runs through the center with the zenith and nadir as its end points; thus zenith and nadir both lie on a line perpendicular to the center point. The center represents not only the contemporary Zuni village, but also the place found in the myth time by Water Skate, the middle of any place, the time of the winter solstice, a person's heart, that person as the center when observing the six directions or offering prayer meal, and so on, all encompassed by the six directions oriented towards that center (Cushing 1897: 7). Which specific center is being referred to at any time depends upon the context; at the same time, "the Center" is a multivocal or condensed symbol that can represent all of these centers at the same time.

Figure 3. Dry painting, fetishes, and wall decoration of the Sword Swallower Fraternity, an esoteric society of the Zuni. By an unknown artist, 1893. Published as Plate 108 in Matilda C. Stevenson's "The Zuni Indians." Courtesy of the Smithsonian Institution, National Anthropological Archives (Neg. No. 2359-C-2).

The annual motion of the sun constitutes a base metaphor in Zuni ceremonialism -- an image of directionality, circularity, and balance that is frequently represented as approximating a semiquartered circle (previously discussed as forming an X and symbolizing the semicardinal directions) plus the zenith and nadir. Thus, the Zuni model is spherical rather than circular; it is based on three rather than two dimensions. The Zunis seldom visually depict the semiquartered circle as a design in itself, however, either in the rock art surrounding the pueblo or in other graphic media.[5]

Instead of constantly rendering this symbol as a design on ritual paraphernalia, making its concrete visual representation a focus, Zunis employ the directionalism it dictates as the basis for patterns of _arrangement_ in ritual. When they constructed the six sacred kivas they aligned them to the solstice directions plus the above and the below (they regard these latter two as symbolically associated with the above and below, not actually aligned to the zenith and nadir). They frequently assign roles in ritual drama so that certain participants represent the six directions (taking on a personality and acting in a manner that is seen as in accordance with these directions). They include symbols on kachina costumes, as well as on the painted bodies of the kachina impersonators, that refer to the directions, painting these symbols in the appropriate directional colors. They position altars in ceremonial rooms, as well as the objects on those altars, so that they are incorporated in a four- or six-directional scheme. They perform ritual actions in a directional pattern, and so on. Thus, the semiquartered circle with the zenith and nadir constitutes a design that operates implicitly, rather than explicitly; as a symbol its basic structure is copied in visually perceivable arrangements and actions. For instance, it is mimed in ritual activity by the gestures of the person who scatters sacred cornmeal to the six directions. It is a metaphor that is acted out, or sketched in the air, but seldom codified in graphic art. Finally, although the model of directionality may be cited as the ideal principle of organization, the Zunis do not always consider it necessary to follow this pattern rigidly in actual practice. The arrangement of the six kivas, for instance, is at best an approximation of the six directions; what is important is the symbolic relationship to those directions, not the actual precision of directional alignments. Further, as alluded to earlier, Zuni potters state that, ideally, pots should be painted in a four-part pattern, but they divide the design field on many pots into threes instead. Perhaps if one recognizes the overall conceptual importance of the pattern, one need not feel strictly obliged to adhere to it.

The allusive yet uncodified property of the base metaphor gives it a dynamic quality, a characteristic permeability or fluidity that adds to its generative power. Indeed, the vast network of symbolic associations to which it gives rise is not amenable to concretization; as one Zuni man said when discussing the various animals, plants, and natural phenomena contained in the six-directional scheme: "It includes everything." But the very generality of this directional metaphor lends it ambiguity -- an ambiguity quite characteristic of the Zuni view of the world. Zuni ritual symbols, whether expressed verbally or visually, are frequently multivalent or multireferential, standing for both themselves and something else _at the same time_; yet all of the meanings are bound together, so that the Zunis say: "They're all the same thing."[6] This system of multiple possible meanings is part of a general pattern in Zuni thinking about the nature of the world, a pattern that is integral to the verbal and visual symbols devised to refer to this world. For example, the various "centers" mentioned earlier, all embedded in the base metaphor of directionality, are linked together by the generalizing concept of the Center; a reference to one can evoke all of the others, yet each has its own distinctive characteristics as well.

For the Zunis, these centers are not simply concrete objects or mental constructs but, rather, powers; although they may be thought of separately, they are also perceived as a single unified power -- the Center -- derived from conflating all of the powers inherent in the activities, objects, relationships, and ideas it represents.[7] But, the boundaries of time and space are also collapsed in the base metaphor, giving it the ability or power to refer to many disparate concepts simultaneously. The Center, a nuclear or condensed symbol that is integral to the conceptual model of directionality, is both a temporal and spatial center; yet it transcends time frames, such as past, present, or future, as well as transcending the limitations of physical space. Significantly, the Zunis' name for their village, their "center," _'itiwana_, is also their word for the winter solstice; one is a center in space, the other a center in time, yet both are "the same thing."

The multireferential quality of Zuni nuclear symbols relates to Zuni principles of aesthetics as well.[8] For instance, "the beautiful" in Zuni thought can also be described as an "aesthetic of accumulation," an elaborate redundancy of symbolism in Zuni sacred and secular environments.[9] Zuni costumes, altars, religious paraphernalia, ceremonial rooms, pottery, wall murals, and the ritual activity itself are all replete with sensory images -- objects, sounds, colors, and so on -- repeated over and over again, all resonating, all dense with meaning, frequently referring to "the same thing." Observations of "secular" activities such as the Zuni Tribal Fair lead to the same conclusions. Secular costumes, exhibit rooms for the fair, the fairground itself, floats in the parade, social dances and music, all are characterized by layers and layers of symbols, colors, and sounds, all bound together metaphorically.

"The same thing" in these instances, the metaphoric base that connects all these facets of Zuni cultural symbolism, is conceptually linked to the motion of the sun along the horizon and to the rain- and snow-bringing winds of the six directions. Indeed, the multiplicity inherent in the Zuni aesthetic of accumulation is also revealed in the generative power of the base metaphor of directionality; it gives rise to a plethora of symbolic associations that extend to include "everything." Further, this concept of multiplicity or accumulation is related to the predominant theme of Zuni religion: the desire for increase, whether it be rain for the crops, success in the hunt, many children, long life, or the sale of much jewelry. Significantly, the central request of most Zuni ritual prayers is that the people may experience increase or prosperity, but this is linked to the more general petition that the Sun Father continue in his appropriate journey and that the rain fall in sufficient abundance; for it is the sun's light, equated by the Zuni with life itself and coupled with water, that provides the subsistence base for all beings. This request is repeated continually, manifested in various but similar ways in Zuni verbal and visual modes of expression.

The significant role of directionality in Zuni world view is also illustrated by their perception of space as inner- rather than outer-directed. According to Zuni mythology, which the Zunis themselves regard as history, the people back in the time of the beginning lived in the fourth underworld, below the surface of the earth. This time of the beginning had no beginning; it simply was, before the time of the emergence. Similarly, Zunis believe that existence after death is not situated in the sky but rather in the Earth Mother, back in the time of the beginning. In the Zuni origin myth, as the people search for the Center after they have emerged from the underworld, they undertake geographical travels that steadily spiral inward until they reach their destination. Their search for knowledge is synonymous with their search for the Center. The vehicle for attaining this knowledge, for finding the Center, is ceremonial activity.

Of further importance to this perspective is the Zuni emphasis on process rather than product; on the unity of all beings in the act of harmonious existence. For the Zunis, the moon and the sun are not objects to be walked on or traveled to but living beings whose light is drawn through the kiva hatchway during certain rituals. This perception of the sun and moon as living beings who enter sacred space at ritually significant times of the year is part of a sequence of ritual activities in which a condensation and intensification of experience occur -- a merging of "inner" and "outer" into one entity. Thus, the Zunis do not view space and the beings who reside there as external to ceremonial life, nor do they regard them as material objects that they can own, control, or overcome. According to this perspective, there are no rigid boundaries between the spiritual and the physical; or, if such boundaries exist at all, they are fluid and permeable. The cosmos is one entity; the beings within it operate according to the principles of continuity and similarity -- principles evident in the unification of inner and outer space in the context of ritual activity.

For the Zunis, this continuity applies to time as well as to space. Although they may introduce a myth as having occurred "a long time ago" or "in the beginning," they do not envision the events of the myth as over and done with, situated at a single point in a linear flow of time; instead, they perceive them as ever-present, informing the here and now. It is this perspective that accounts for the "presentness" of the beings of myth and folktale in Zuni life.

This view of time and space is perhaps closer to that of relativist physics, in which space and time exist as a single continuum that is relative to the observer, than it is to the linear perspective of the average member of Western society (Edelman 1973:305). For the Zunis, time is cyclical, apparent in the orderly and regular motions and "returns" of the sun, moon, and stars, and both time and space are organic, continuous entities. One may say that time for them is reversible; past, present, and future are coexistent. They believe that when people die four times, they return to the time and place of the myth, completing the human cycle of reincarnation; yet they may return to the present in another form if they so desire. Further, in the context of sacred performances, the gods return again and again. The kachina dancers in the plaza do not represent the gods, they become the gods, and the time of the myth is one with the present. The efficacy of ritual activity is the result of the merging of the here and now with the myth time and space. Thus one might describe the Zuni view of the universe and its dimensional aspects as a unified and orderly phenomenon, symbolized by the directional model that links a complex web of associations, extending to include "everything." It may well be that the power of myth, the power of certain visual images, and the power of the ritual process inhere in this world view as well.

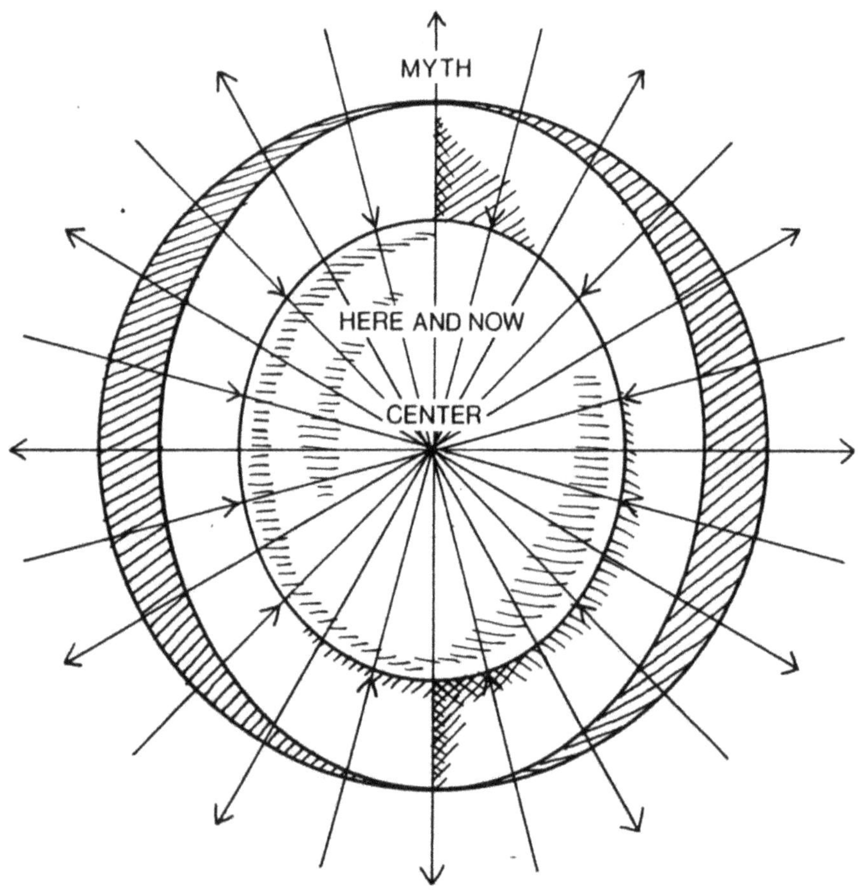

Figure 4. Schematic diagram of the dialectic between the "here and now" and the myth time. The concentric spheres revolve about a common center or radius (the Zuni *'itiwana*). Drawing by Murray Callahan.

Although I have discussed Zuni spatial and temporal perceptions in terms such as the "merging," collapsing," and "coalescing" of boundaries between the here and now and myth time and space, I do not mean to imply that this results in a permanent condition so that the two states (time and space; myth and "here and now") are unified in every aspect from that moment on. Rather, the relationship between the two is best expressed as a dialectic interaction made possible by the fluid boundary that exists between the two. Figure 4 illustrates this movement back and forth across a permeable boundary; the here and now validates the myth and the myth validates the here and now. In this diagram the two states are depicted in three-dimensional space -- they are represented by concentric spheres. The Center from which both spheres have been generated is the nuclear symbol discussed earlier: 'itiwana, the center of all time and space. The "here and now" sphere is congruent with the model of the world depicted in Figure 1, but this directional scheme does not include the myth time and space that extend beyond the limitations of the present. Furthermore, the volume of neither sphere portrayed in Figure 4 is constant. Given certain circumstances, either sphere may expand or contract such that the conceptual distance between the two becomes greater or less. For instance, the rituals that occur near the time of the winter solstice (such as the "coming of the gods" to Zuni), the most critical time of the year, draw human time closer to myth time so as to intensify the relationship between the two (depicted schematically, this would mean that the inner sphere in Figure 4 expands and moves toward the outer sphere). The recitation of ritual prayers and reiteration of myths during this period contribute to this intensification, acting as catalysts in a world renewal rite, bringing the beings and events of myth time close to hand.[10]

NOTES

A shorter form of this article was presented at the symposium, "Ethno- and Archaeoastronomy in the Americas," during the 46th International Congress of Americanists in Amsterdam, Holland in July 1988. The forum was organized by Edmundo Magaña, Elizabeth Reichel, and Anthony F. Aveni. For an expanded version of the topic discussed in this article, see Chapter 3 ("Zuni Cosmology and Cultural Symbolism") of Young 1988.

I thank the Tribal Council of the Pueblo of Zuni for its support and am greatly indebted to my Zuni friends for their hospitality during my periods of fieldwork at Zuni -- special thanks go to the late Andrew Napetcha, Gus Panteah, Alex Seowtewa, Arlen and Virina Sheyka, and Frank Vacit. I am grateful to Murray Callahan and Snowden Hodges for permission to use their splendid drawings in this article, and to the National Anthropological Archives of the Smithsonian Institution for allowing me to include one of their historic photographs.

[1]Although there is no longer a sun priest at Zuni, this duty has been taken over by another official.

[2] It is noteworthy that the number four is the sacred or pattern number of many, although by no means all, Native American peoples. Thus, the importance of the number four for the Zunis is not peculiar to them but part of a broader context.

[3]This description of the semicardinal directions as the ritual directions and counterclockwise as the ritual order is a generalization. I am grateful to Barbara and Dennis Tedlock for pointing out to me that at Zuni the use of cardinal versus semicardinal directions, as well as the use of clockwise versus counterclockwise, depends on the particular ceremonial context. In his unpublished notes, J. P. Harrington gives "archaic Zuni words" used for the directions in Zuni ceremonies. He translates these directional words as the semicardinal directions, that is, "northeast, northwest, southwest, southeast," and states that they refer to the solstice extremes. See J. P. Harrington, n.d.

[4]When cardinal rather than semicardinal directions are employed as the directional model, the following associations are made: yellow mountain lion/north, blue bear/west, red badger/south, and white wolf/east. See Cushing 1883: 16-16, 24-25. Each of the species of prey animals is again divided into six varieties, according to color. Thus, for example, although the mountain lion is primarily associated with the north and the color yellow, there is a mountain lion representative for each of the other directions/colors as well.

[5]There is a Zuni symbol used in certain contexts to stand for the six-part directional scheme, the hepakinne, that has been described as a sunflower as well as corn and squash blossoms. See Bunzel 1972: 92, 95; Stevenson 1904: 134. Painted on Zuni water jars and on the masks of the Salimobia and Shitsukia kachinas, the hepakinne is a hexagonal design inside a circle, generally with each division painted in one of the colors associated with the six directions. Among other meanings, Zunis say that it represents the "differently colored lightning"

associated with those directions (Bunzel1932: 923). Nevertheless, my point is that, although it may be used occasionally, the four- or six-part design is not so much a prevalent symbol in Zuni graphic art as a prevalent pattern of arrangement and association.

[6]Such multivalent or multireferential symbols have been described by Victor Turner as nuclear or condensed symbols. See especially V. Turner 1967: 295, 298. See also Farrer 1977, 1980; Farrer and Second, 1981; Munn 1973: 171-73.

[7]Victor Turner (1967: 298), discusses the power of conflation inherent in a nuclear symbol.

[8]For a complete description of the Zuni system of aesthetics, see B. Tedlock 1984: 246-65.

[9] I first heard the term "aesthetic of accumulation" in an informal colloquium on the home altars of Mexican American women presented by Kay Turner, then a graduate student in the Folklore Program at the University of Texas at Austin. In subsequent discussion we agreed that the same sort of principle of elaborate redundancy seemed to operate at Zuni. Although she does not specifically use the term "aesthetic of accumulation" in her writings, Turner's articles offer an excellent portrait of the principle of elaboration in Mexican American folk art. See especially Kay Turner 1982: 309-26.

[10] I am grateful to Dell Hymes for his comments and suggestions that have served as catalysts in the formation of this discussion of the dialectic interaction between the myth time and the "here and now."

REFERENCES CITED

Benedict, Ruth

1935 Zuni Mythology. 2 vols. Columbia University Contributions to Anthropology, no. 21. New York: Columbia University Press.

Bunzel, Ruth L.

1932 Introduction to Zuñi Ceremonialism, Zuñi Origin Myths, Zuñi Ritual Poetry, Zuñi Katcinas: An Analytical Study. In Forty-Seventh Annual Report of the Bureau of American Ethnology for the Years 1929-1930, pp. 467-1086. Washington, D.C.: Government Printing Office and Smithsonian Institution.

1972 The Pueblo Potter: A Study of Creative Imagination in Primitive Art. Reprint of 1929 edition. New York: Dover Publications.

Cushing, Frank H.

1883 Zuni Fetiches. In Second Annual Report of the Bureau of American Ethnology for the Years 1880-1881, pp. 3-45. Washington, D.C.: Government Printing Office and Smithsonian Institution.

1896 Outlines of Zuñi Creation Myths. In Thirteenth Annual Report of the Bureau of American Ethnology for the Years 1891-1892, pp. 321-447. Washington, D.C.: Government Printing Office and Smithsonian Institution.

1897 Discussion and Remarks on Shamanism. Proceedings of the American Philosophical Society 36: 1-14.

Edelman, Prewitt
1973 Working Back: The New Physics and Pueblo Mythology. Southwest Review 58: 302-06.

Farrer, Claire
1977 Play and Inter-Ethnic Communication: A Practical Ethnography of the Mescalero Apache. Ph.D. dissertation, University of Texas at Austin.

1980 "Singing for Life": The Mescalero Apache Girls' Puberty Ceremony. In Southwestern Indian Ritual Drama, edited by Charlotte J. Frisbie, pp. 125-59. Albuquerque: University of New Mexico Press.

Farrer, Claire, and Bernard Second
1981 Living the Sky: Aspects of Mescalero Apache Ethnoastronomy. In Archaeoastronomy in the Americas, edited by Ray A. Williamson, pp. 137-50. Los Altos,Calif., and College Park, Md.: Ballena Press and Center for Archaeoastronomy.

Harrington, John P.
n.d. Unpublished manuscripts and papers pertaining to fieldwork at Zuni. National Anthropological Archives, Smithsonian Institution.

Hymes, Dell
1980 Particle, Pause and Pattern in American Indian Verse. American Indian Culture and Research Journal 4 (4): 7-51.

Munn, Nancy
1973 Walbiri Iconography: Graphic Representation and Cultural Symbolism in a Central Australian Society. Ithaca and London: Cornell University Press.

Parsons, Elsie C.
1917 Notes on Zuñi. Memoirs of the American Anthropological Association 4 (3-4): 151-327.

1923 The Origin Myth of Zuñi. Journal of American Folk-Lore 39: 135-62.

Stevenson, Matilda C.
1904 The Zuñi Indians: Their Mythology, Esoteric Fraternities, and Ceremonies. In Twenty-Third Annual Report of the Bureau of American Ethnology for the Years 1901-1902, pp. 3-634. Washington, D.C.: Government Printing Office and Smithsonian Institution.

1915 Ethnobotany of the Zuñi Indians. In Thirtieth Annual Report of the Bureau of American Ethnology for the Years 1908-1909, pp. 31-102. Washington, D.C.: Government Printing Office.

Tedlock, Barbara
1984 The Beautiful and the Dangerous: Zuni Ritual and Cosmology as an Aesthetic System. Conjunctions: Bi-annual Volumes of New Writing 6: 246-65.

Tedlock, Dennis
1972 Trans. Finding the Center: Narrative Poetry of the Zuni Indians, by Andrew Peynetsa and Walter Sanchez. New York: Dial Press.

1979 Zuni Religion and World View. In Handbook of North American Indians, Southwest, Vol. 9, edited by Alfonso Ortiz, pp. 499-508. Washington, D.C.: Government Printing Office.

Turner, Kay
1982 Mexican American Home Altars: Towards Their Interpretation. Aztlan--International Journal of Chicano Studies Research 13: 309-26.

Turner, Victor
1967 The Forest of Symbols. Ithaca and London: Cornell University Press.

Young, M. Jane
1988 Signs from the Ancestors: Zuni Cultural Symbolism and Perceptions of Rock Art. Albuquerque: University of New Mexico Press.

ASTRONOMY AND RITUAL:
The Rhythm of the Sacred Calendar in the U.S. Southwest

Michael Zeilik
Department of Physics and Astronomy
The University of New Mexico
Albuquerque, New Mexico 87131
U.S.A.

Introduction

Archaeoastronomy in the U.S. Southwest aims to infer the content and the impact of astronomy in these prehistoric societies. For this study, the Southwest has a rich historical advantage--the Pueblo people living there who have a cultural trajectory from prehistory. That path runs especially true with the historic Pueblo people directly from their immediate protohistoric ancestors and on to their Anasazi (and Mogollon) ancestors who lived throughout the northern Southwest in the broad Anasazi region (now known as the Four Corners area).

This paper focuses on the astronomical cultural heritage of the present-day Pueblos revealed in the ethnographic record and also aims to delineate astronomical tradition of the earlier Anasazi. Here the past comes alive when the Shalako ceremony takes place at Zuni, regulated by lunar observations. I can easily imagine a similar ceremony--different in detail but with similar ritual intentions--in a prehistoric pueblo. That mental image contains the essence of the oft-maligned argument by ethnographic analogy that I will use in a conservative way: as a culturally-appropriate source of hypotheses for testing in a prehistoric context (Binford, 1972; Reyman, 1975). I presume that those Puebloan astronomical concepts that are most frequent within the diversity of the historic Pueblos mark cultural traditions that are the deepest in time. Specific cultural discontinuities certainly <u>did</u> occur between prehistory and now--but aspects essential to survival and adaptation to the high desert stayed intact or were rediscovered and revived. In particular, the sacred and planting calendars, kept by observations of the sun and the moon, generate the rhythms of the life of the historic Pueblos--an essential integration of astronomy, agriculture, and ritual. The same was likely the astronomical practice of the Anasazi.

Note that to generate testable hypotheses by ethnographic analogy does <u>not</u> mean or require a direct continuity in time. We know that the Anasazi migrated from Chaco Canyon and Mesa Verde to settle in new areas, notably the Rio Grande and Chama river valleys (Cordell, 1984:328-337). When the Spanish arrived in the region, they brought with them European diseases that prompted the decline of the Pueblo population and the abandonment of villages (Simmons, 1979; Schroeder, 1979). Such episodes could result in the loss of astronomical knowledge, but it could be reinvented. The key point is that the rediscovery takes place in a similar cultural context for a similar adaptive purpose. For example, only two naked-eye techniques of astronomy can provide reliable and accurate seasonal calendars---and the Pueblo people discovered and used <u>both</u> of them.

To explore the Pueblo cultural context for astronomical practices requires an examination of the interior of Pueblo religion and daily life, as they connect to astronomical observations and integrate them into a cosmic world view. The Pueblo people can best say this themselves, as Don Talayesva, a Hopi from Third Mesa, states in Sun Priest (Simmons, 1942; 58-59):

> Another important business was to keep track of the time or the seasons of the year by watching the points on the horizon where the sun rose and set each day. Old Talasemptewa, who was almost blind, would sit out on the housetop of the special Sun Clan house and watch the sun's progress toward its summer home. He untied a knot in a string for each day. When the sun arose at certain mesa peaks, he passed the word around that it was time to plant sweet corn, ordinary corn, string beans, melons, squash, lima beans, and other seeds. On a certain date he would announce that it was too late for any more planting. The old people said that there were proper times for planting, harvesting, and hunting, for ceremonies, weddings, and many other activities. In order to know these dates it was necessary to keep close watch on the sun's movements.

The Pueblo people perceive a world in which the correct rituals must take place at the proper times to keep in harmony with the cosmos--and those "proper times" are set by astronomical observations.

Astronomy in the Historic Pueblos

The ethnographers who visited the Pueblos before the economic encroachment of Anglo society in the 1930s left detailed records of many aspects of Pueblo life. So many and so detailed are they that the astronomical unity of the calendar gets lost--in part, because these people rarely focused on the astronomy as an observational practice. The published and unpublished information provides a rich but scattered resource to delve into Pueblo astronomy.

In general, we should distinguish between astronomical purposes and the astronomical practices that derive from the purposes. In the Pueblo Southwest context, astronomy serves the purposes of establishing and validating (1) sacred directions and cosmic patterns, (2) cosmic mythology, (3) certain ritual sites and shrines, (4) the ritual and planting calendar, and (5) times for hunting and gathering. These desired ends prompted the practices of horizon calendars, light and shadow markers, and lunar phase counts for tracking the calendar. The main task of the calendar watch centers on methods to forecast festival dates by anticipatory observations (McCluskey, 1982; Zeilik, 1985a, 1986a). Pueblo ceremonies must be announced ahead of time so that ritual preparations can be properly carried out, and the meshing of solar/lunar ceremonies worked out by the proper intercalation (McCluskey, 1977). The ceremonial cycle typically spans one year, and solar and lunar observations conducted by the appropriate religious officials set the timing for the rituals that are laid out in a sequence, where the end of one ceremony marks the start of the next.

Anticipation provides a culturally-correct way to understand the terms "precision" and "accurate," when combined with historical information about festival dates as has been done by McCluskey (1977) for Hopi and Lyon (1985) for Shalako at Zuni. Historically, sun priests at Hopi and Zuni were able to fix solstice ceremonies to within the astronomically correct day (or two) by making anticipatory observations about two weeks ahead of time (Zeilik, 1985a). In contrast, lunar ceremonies appear to be moveable festivals, with a scheduling decision under the power of the person responsible for setting the date (Frigout, 1979; Sekaquaptewa, 1983; Zeilik, 1986a). The date of a particular lunar phase (such as full) may have been established to within a day (or two) of the astronomically correct phase (Bunzel, 1932b:702-703).

An emphasis on calendric anticipation switches our attention from the positions to rate of change of position (angular speed) of the sun in keeping a seasonal watch. This mental realignment should not, however, direct us to overlook the importance of direction to the Pueblo sense of sacred space. The special directions are usually the four solstitial ones (winter and summer rising and setting), as well as the zenith and nadir, such as embodied in the placement of corn ears in the Chief kiva during the Hopi Powamu ceremony (Stephen, 1936:228, figure 140). Special colors are also linked with each of the sacred directions; animals too, are associated with each direction and color (Parsons, 1939). Such sacred directions may have been the impetus for the design of buildings with special orientations in historic and prehistoric times. The point here is that cycles such as the sun's seasonal swing along the horizon involve motions to achieve the extreme positions. The Pueblos embodied both in their astronomical practices: the directions tend to be cosmic; the motions, calendrical.

Pueblo Sun Watching

How were the calendrical observations made? For the sun, a horizon calendar was most commonly used, and we have detailed maps for a few of them (Stephen, 1936: maps 4 and 12). Astronomically, such horizons need to be fairly distant (a few kilometers at least) with some, but not necessarily dramatic, relief. The sun watcher, whom I will call the "Sun Priest," also goes to a special sunwatching station, which is typically not marked by rock art (Young, 1986a; Zeilik, 1985d). The location may simply be remembered, as part of the esoteric knowledge of the Sun Priest's office.

Note that if the horizon is distant--ten or so kilometers--the choice of the standing spot is not crucial if it is within or close to a pueblo. Among the historic Pueblos, the sun priest rarely travels very far to make his observations--from the roof at Walpi (Stephen, 1936: 62), near the edge of the mesa for Hano (Yava, 1978: 72), or a few kilometers east from Zuni (Halona) to the deserted village of Matsakya (Stevenson, 1904: 118).

The second calendrical technique involves the use of light and shadow cast through windows/portals against a wall with markers (Lange, 1959:56); this method yields more data in the form of archaeological remains. The famous passage written by Cushing (1979; 117) indicates that markers of some kind on the wall were common features of dwellings: "...for many are the houses in Zuni where scores on their walls or ancient plates imbedded therein, while opposite, a convenient window or small port-hole lets in the light of the rising sun..." The diaries of John G. Bourke tell us what was really going on (Bourke, 19 November 1881):

> After breakfast, Cushing, Pedro Pino [the Governor of Zuni] and myself went to the upper story of one of the highest houses on the Eastern side of the Pueblo; here in the West wall was an old blue china plate fixed there, so the head of the house said, in the time of the Spaniards, to conceal a painting of the Sun, which faced a small rectangular aperture in the eastern wall. When the sun shone through the aperture farthest to the North, Spring had come and the season of planting had arrived; the more Southerly aperture allowed the rays of the Sun to fall upon the center of the plate (in ancient times upon the face of the sun picture) about the period of the Autumnal equinox--and when the light struck a certain point in the wall, it was the time of the Winter Solstice.

This statement reveals the practice of establishing solar calendars at Zuni at a time before Anglo culture had much impact. Sunlight hits the Sun face during harvesting season; planting and the winter solstice were also noted. And we also know that at the same time, Pekwin, the Sun Priest (or "Cacique of the Sun", as Bourke called him) conducted a horizon watch (Bourke, 19 November 1881) as well as keeping a wall calendar in his own house (according to his brother). The half of the wall calendar that marked the time from spring planting to the summer solstice consisted of a horizontal line of scratches onto which sunlight fell through an east-facing window. The half of the year before the winter solstice was marked by strings of abalone shells hanging on the wall.

Another light and shadow process also appears at Zuni, where Pekwin visited a "small post of petrified or silicified wood" (Fewkes, 1891:4). Fewkes states that this post "...is in certain particulars a gnomon"--an implication that the position of its shadows was used as a marker at sunrise. Bourke (19 Nov. 1881) clarifies the situation a bit, when he writes that the stone pillar was a place for Pekwin to make sacrifices and watch the sun. Bourke and Cushing go with the Zuni governor, Pedro Pino, to

> the Eastern boundary of the Pueblo to an open field in the center of which was a vertical, four sided piece of petrified wood, put there, Pedro said, a long, long while ago. In line with it away off to the East, six or seven miles on the crest of a mesa could be discerned another monument, pointed out by Pedro.

I infer that Pekwin used the pillar in the field to position himself in the right alignment with the pillar on the mesa.

I find these Zuni examples especially curious because we are told that Pekwin keeps <u>both</u> a horizon watch (Stevenson, 1904:109, 117-118) and also a wall calendar. What was the cultural reason for this redundancy? Did Pekwin use both to anticipate and confirm important times? For instance, one major sequence of the wall calendar includes the time of planting, winter solstice and summer solstice (Bourke, 3 September, 1881). Pekwin's horizon calendar involves similar time sequences. Did these two parallel calendars both support the degree of precision that Pekwin was supposed to achieve? Bourke (25 November 1881) states that

> Cushing took me to call upon the Priest of the Sun. This official, in the different Pueblos, is supreme during time of peace, that is to say his edicts can only be revoked by a council of all the Caciques. The present incumbent of the office in Zuni is a very young man and has but recently entered upon his duties. I have noted elsewhere [20 November 1881] that <u>by a mistake in his orientation</u> he ante-dated the festival of the Winter Solstice by some (15) or eighteen (18) days, a mistake very summarily corrected by the Council of the Caciques [emphasis mine].

Note that this passage tells us that other religious officials kept watch on Pekwin's work, and could track the calendar as well as he to insure that "correct" times were announced. It tells us that Pekwin's mistake in anticipatory observations was far too large to be acceptable to the council.

I have discussed Zuni in some detail as a specific example of pan-Pueblo practices and as a case where we have substantial information -- most of it consistent! -- predating the 20th century. The practices may best reflect the survival of an astronomical tradition that traces back to prehistoric times.

In summary, the ethnographic evidence about the historic Pueblo solar calendar indicates:

(1) attention to a one-year <u>seasonal cycle</u> with the winter solstice usually as the "middle";

(2) a <u>fixed sequence</u> of ceremonies, in which the completion of one sets the stage for the next;

(3) an <u>anticipation</u> of ritual times that are forecast early enough so that proper preparations can make the ceremonies effective;

(4) <u>astronomical observations</u> that set the calendrical schedule.

The winter solstice, planting season (mid-April up to the summer solstice as for the Hopi--see Parsons, 1925a; Beaglehole, 1937; Forde, 1931; Titiev, 1938), and the summer solstice are the rank order of importance for seasonal dates. The pueblos rarely mark the equinoxes as such, though they do correlate roughly with planting and harvesting. See Ortiz, 1969:104, for explicit comments about the importance of the equinoxes to the Tewa, which marked the seasonal transfer of power between the summer and winter people. I will define <u>calendrical site</u> to mean one that could be used to forecast and confirm these dates with a degree of accuracy of a day or two.

Anasazi Solar Horizon Calendars

Because they are so prosaic, horizon calendars have received little attention--even though we have the most and best ethnographic information about them! They are hard to substantiate archaeologically, for it is difficult to ascertain where a sun priest might have stood to make his observations (Williamson <u>et al.</u>, 1977). By ethnographic analogy, one can search for a spot in or close to the archaeological site of a village that provides a view of the horizon to see if a calendar can be kept in this fashion.

Anasazi sites that have been tested for horizon watching include: Penasco Blanco (Williamson, Fisher, and O'Flynn, 1977; Zeilik, field notes), Wijiji (Williamson, Fisher and O'Flynn, 1977; Williamson,

1983; Zeilik and Elston, 1983), Fajada Butte (Williamson, Fisher, and O'Flynn, 1977), and Pueblo Bonito (Zeilik, 1986b)--all in Chaco Canyon; and Tsiping Pueblo in the Chama valley of New Mexico (Zeilik, 1984a), Tsankawi (Zeilik, field notes) on the Pajarito Plateu, and Kuaua along the Rio Grande (Zeilik and Lowther, 1988 and field notes). The Chacoan Great House sites range in dates from 900 to 1100 A.D. (those on Fajada Butte date to the 1200's); Tsiping is a later occupation, from 1250 to 1450 A.D., and a probable prehistoric Tewa site, as is Tsankawi (now part of the Bandelier National Monument), which may have been occupied up to the late 15th century. Kuaua was occupied from 1300 A.D. into historic times and may have been visited by the Spanish (Dutton, 1963:3-18); let's look at this one in detail for an example of a potential horizon calendar.

Kuaua (Tiwa for "always green") is situated on the west bank of the Rio Grande, about 30 km north of Albuquerque. It was made of coursed adobe and contained over 1200 rooms, with 3 plazas and 6 kivas (Vierra, 1987). Kuaua contains, in its square kiva, famous ceremonial murals (Dutton, 1963). Some of these sacred figures depict the Sun Father (based on ethnographic interpretations) and relate to ceremonies at the winter and summer solstice, as well as general symbolism related to fertility. At the site today (part of Coronado State Monument), the tree line of the Rio Grande makes observations difficult from the ground. M. Lowther and I therefore observed from the roof of the Museum, which would mimic observations from the second story of the pueblo. The Sandia and Ortiz Mountains to the east provide a clear horizon profile at sunrise from the equinoxes to the winter solstice (Zeilik and Lowther, 1988). We know that at Santa Ana Pueblo, the Sandia Mountains are used to track the approach of the winter solstice (White, 1942:92,103), for which Masewi (the war priest and one of the Twin War Gods) begins the watch in September, just the time when the sun begins to climb up the slope of Sandia. Dutton (1963:205) states that some of the Kuaua people may have joined those at Santa Ana.

Our firmest ethnographic information about calendrical sun watching involves horizon anticipatory sequences. Hence, each Anazasi site should be checked carefully for a useful horizon track--or lack of one! If horizon watching is not workable for certain intervals, we may then infer that light and shadow techniques might have provided an alternative procedure.

Light and Shadow Calendric Markers

Light and shadow phenomena draw our attention by their visual appeal, especially through the medium of photography. The three-slab site and other sites at Fajada Butte (Sofaer and Sinclair, 1987) and the petroglyph sites in the Petrified Forest (Preston and Preston, 1985, 1987) have sparked public attention to this one aspect of Southwestern archaeoastronomy. The petroglyph site near Holly House in Hovenweep is also visually stunning (Williamson and Young, 1979; Williamson, 1979, 1981, 1984), but has a lower visibility. Other light and shadow sites associated with window/portals in buildings include the corner openings at Pueblo Bonito (Reyman, 1976; Zeilik, 1986b); the northeast opening in Casa Rinconada in Chaco (Williamson, 1982, 1987; Zeilik, 1984b); and openings in add-on rooms at Hovenweep Castle, Unit Type House, and Cajon Ruins at Hovenweep (Zeilik, 1987).

Are such sites intentional and accurate calendrical markers? To make such a case requires, in my opinion, that the site serve to anticipate the solstices with an accuracy of a day or two and to keep track of a planting calendar. This requirement brings up the issue of practical resolving power--the ability of an observer to interact with a site to mark reliably one-day differences in shadow positions. Note that an emphasis on anticipation shifts our attention from the positions of the light/shadows to the rate of change of position on a surface--linear speed.

The linear speed along the surface per day, as the shadow mimics the seasonal motion of the sun, needs to be large enough to observe reliably. Also limiting the usefulness is the sharpness of the shadow edges. When the light source has a finite size, such as the sun (which subtends 1/2 degree in the sky), a shadow has a fuzzy edge, even if the shadow-casting edge is smooth. Given the eye's perceptive ability for edge contrast, I suspect that a linear motion of at least 0.5 to 1 cm/day is needed for reliable observations with a one-day resolving power. To analyze the site correctly requires that we interact with it in the same way as a prehistoric observer might. So, the roughness of the shadow-casting edges (which will increase the fuzziness of the shadow), the texture and color of the surface on which the shadows fall, the sharpness of petroglyphs/pictographs on the surface, and the viewing position and angle all enter into the character of the observations. These are all specific to the site, so each site should be judged based on these details of the site's geometry.

Light/shadow prehistoric sites can be divided into two groups: interior (buildings) and exterior (rock formations with art). Here I deal with two examples of each.

At Hovenweep, Williamson, Fisher, and O'Flynn (1977) investigated a number of tower structures. Williamson (1981) elaborated on Hovenweep Castle and the Cajon Group ruins. These places and the Unit Type House have add-on rooms that contain portals apparently positioned to let in light for calendrical purposes. (Some portals are vents; others provide sights to other buildings in the canyons--not every one has an astronomical purpose.) Cajon lies about 10 km south of Hovenweep Castle and Unit Type house, which are within a kilometer of each other. Zeilik (1987 and field notes) re-examined these sites and found that all could be used in an anticipatory fashion and that the average speeds of the light beams on the walls ranged from about one to a few centimeters per day. Photos taken by Fewkes (1919) just prior to 1900 indicate that for Unit Type House and Hovenweep Castle, the portals were intact before any stabilization. Hence, they were originally an integral part of the structures, although we are not sure to what extent the insides of the openings were plastered and so would restrict the view compared with today.

Cajon and Hovenweep Castle work calendrically at sunset; Unit Type House in early morning (about 1/2 hour after local sunrise because of a nearby ridge that blocks the sunlight). In all three places, the horizon view does not lend itself to work well as a horizon calendar. In contrast, the play of light within the rooms against the walls of each can permit one to anticipate the solstices, confirm them, and keep track of a planting calendar. I judge them, therefore, to be the best examples we have found to date of intentional calendrical structures, in a large part because of the integrity of the structures.

The three-slab site atop Fajada Butte in Chaco has the widest renown among the Anasazi exterior sites (Sofaer, Zinser, and Sinclair, 1979). Sofaer and Sinclair (1987) have improved the knowledge of the ethnographic context and also have expanded the view to other petroglyph sites on the butte, which they claim mark local solar noon and the seasons. Sofaer and colleagues find particularly attractive the purported integration of the sites for sun/moon and season/noon observations. Newman <u>et al</u>. (1982) infer that the three-slab site is likely a natural rockfall, rather than a "construct." I have reevaluated the three-slab site in some detail (Zeilik, 1985b, 1985c, 1986c, 1986d; see also Carlson, 1987; Sofaer and Sinclair, 1986a, 1986b) and will not repeat those comments here.

In my judgement, the three-slab site fits much better with the attributes of historic sun shrines (Fewkes, 1906) than calendrical observational places. If so, the butte may have been a sacred--even cosmological--place, but I think that its astronomical use for calendrical keeping was much less likely its purpose. That is, the sunlight motions at the three-slab site act as a <u>seasonal hierophany</u> that contains calendrical information but has limited practical resolving power and so is not useful for forecasting solstitial dates accurately (see Hedges, 1985:27 for a similar conclusion).

The Holly site at Hovenweep presents a dramatic display at (and before) the summer solstice. Then, about a 1/2 hour after local sunrise, two shafts of light move horizontally across a petroglyph panel that includes spiral and circle shapes. Only about 7 minutes pass from the first appearance of the beams and their merging near a spiral. Williamson and Young (1979) have noted symbols on the panel that may relate to the Twin War gods (Venus as Morning and Evening star) and a snake/serpent that may be a water symbol. Williamson (1984:93-103) has concluded that the site may have been used to keep a planting calendar similar to that at Hopi--especially pertinent to the farming area in which it is located. He also suggests that the change in the positions of the light shafts would allow a practiced observer to anticipate the summer solstice to "...within three or four days". (In contrast, the summer solstice portal in Hovenweep Castle would allow a trained observer to predict the summer solstice to within a day or two.)

In summary about exterior sites: I concur with Young (1986) and Schaafsma (1985:264-266) that some Anasazi rock art served to interact with light and shadow. However, I think that a better case needs to be made that these interactions work <u>calendrically</u> in the appropriate cultural context. No ethnographic analogy has been found yet for exterior sites whose light/shadow interacts with rock art for calendrical purposes. (Of course, the physical principle is the same as that for light entering the windows of buildings.)

Historic Pueblo Lunar Observations

Ethnographic information shows that some Pueblos paid close attention to the phases of the moon and used them as a timing device (among other functions, such as weather prediction) for some festival dates--most notably Shalako at Zuni (Zeilik, 1986a) and Powamu at Hopi

(McCluskey, 1977). The regular phase cycle serves as a nightly marker that can be used to anticipate events that coincide with a specific phase. For example, at Zuni, the head of the Long Horn katchina group had to anticipate full moons (for preparation and planting of prayer feathers by the impersonators of the Shalako) by two days (Bunzel, 1932c:963). He could have done so by spotting the first visible crescent after new moon and then counting for 10 days (Zeilik, 1986a).

Because the Pueblos had a reliable seasonal calendar (based on solar observations), the use of the moon to schedule ceremonies brings up the ancient problem of synchronizing the solar and lunar calendars. This intercalation issue clearly arises at Zuni in the coordination of the sequence of Shalako, the Winter Solstice ceremony, and the New Fire ceremony that marks the start of a new ritual year. Pekwin sets the date for the Winter Solstice ceremony by observations of the sun; Sayatasha (the head of the Longhorn Kachina group) establishes the Shalako schedule by observing the phases of the moon and has the final word in the decision for the date of this ceremony (Bunzel, 1932c). Sayatasha consults with Pekwin to ensure that Shalako does not conflict with the Winter Solstice festival.

Meanwhile, we are also informed that Pekwin is supposed to arrange for the Winter Solstice ceremony to occur at the full moon in December (Bunzel, 1932a:534; see also Tedlock, 1983)! Reyman (1980) has emphasized how troublesome this scheduling requirement will be for Pekwin, for even a three-day range, at best sets the full moon/winter solstice coincidence to happen only every three years. However, Pekwin does <u>not</u> control the date of Shalako, which <u>is</u> based on lunar phases. The 49-day countdown to Shaklo begins on the day of a full moon (Bunzel, 1932a; 523), and that is directed by Sayatasha. Another articulation of the sun and moon occurs at the Winter Solstice ceremony, for the first full moon that follows begins the count of 10 full moons that is the first stage of the timing for Shalako. Overall, I do not see how Sayatasha and Pekwin kept the timings in line.

One hint comes from the sequence of Zuni months, in which the first six are named and the next six nameless (Cushing; 1920:154) or repeated in names (Stevenson, 1904:108). Parsons (1917:300-301) suggested that the unnamed November moon, in particular, provides the elasticity to reconcile the Shalako and Winter Solstice ceremonies. Shalako dates (Lyon, 1985) need to be compared carefully to Winter Solstice celebratory dates to illuminate this issue (as McCluskey, 1977, has done for the Hopi).

The sun and moon watching practices of the Pueblos invoke the intercalation problem. The months began with the observation of the new (first visible crescent) moon; the seasonal cycle by sun watching. The months tended to be named after key ceremonies that occurred within them (especially at Hopi; see Ellis, 1975:65-68 for a summary) or seasonal by events (see Harrington, 1916:63-66 for the Tewa month names and a comparison to those at Jemez and Zuni). The Pueblos that tally and name months usually have 12 or 13 in a seasonal year (see Cope, 1919:151-153; Spier, 1955). I suspect those with a fixed 12 months have adopted, at least in part, a European calendar. Those that have 13 months (or more) probably are the closest to the aboriginal calendar, for those allow intercalation by a "short month" technique. One Hopi, Abbott Sekaquaptewa (1983), has written in a

Third Mesa newspaper that, for the case of the winter solstice ceremony and the next new moon (Pamuya), the adjustment involves "the elimination of one or a couple of the four day cycles after the final rites of the winter solstice ceremony." This procedure speeds up the arrival of the next lunar cycle and "allows a continuity of the normal ceremonial and spring planting times."

McCluskey (1977) has analyzed the actual dates for the Hopi winter bean ceremony (Powamu), where the full moon after the close of the winter solstice ceremony signals the planting of beans in kivas (Parsons, 1939:505; Stephen, 1936:162). Now, each year a given phase of the moon arrives 10.9 days earlier, so that after three years, the moon phases are 32.7 days "ahead" of the sun. A simple intercalation scheme would occur every 2.71 years. McCluskey (1977) finds this periodicity in the Powamu dates. The inference is that one "month" is not counted every 2.71 years; it may be a very short interval. Unfortunately, similar evidence needs to be gathered for other Pueblos, and no general, consistent scheme of intercalation appears in historic times.

Lunar calendars kept by phases (with the first visible crescent starting the month) present a serious archaeological problem: What would be the material remains? We have two hints. At San Ildefonso, Stevenson (n.d.) reported that the moon watcher kept a tally for the year by notching the edge of a flat stone on which was drawn a moon face. These stones were then deposited in a ceremonial chamber. At Zuni and Hopi, we have evidence of the use of calendar sticks to track the months. Fewkes (1892:151) described the one at Hopi, and Alexander Marshack, at my suggestion, found the same stick in the Smithsonian Institution. The Hopi stick has a total of 28 grooves, which could be used to count or record a month of phases if the times of invisibility are not counted. The Zuni stick is described and drawn by Bourke (21 November 1881 based on information acquired in a meeting with the Zuni Nayuchi, who was Head Bow Priest at the time). Bourke implies that it was used to tally a year, starting with "...the moon of the Fire Festival, or winter solstice". By Fire Festival, Bourke means the New Fire ceremony, which takes place right after the Zuni winter solstice ceremony. Fewkes (1892: 151) states that he has seen calendar sticks at Walpi and Hano. He describes them as follows:

> These sticks are about a foot or a foot and a half long, and are divided into two parts, one section being round, the other flattened on one side. The round section is girt by fifteen shallow parallel grooves, and occupies about a third of the whole length of the stick. The remaining two thirds of the stick has a number of parallel grooves or notches cut upon the flattened surface. Five of the latter grooves, which are situated at equal distances, are deeper than the remaining, and between each pair there are four smaller parallel grooves arranged at equal distances. The space in which these grooves are cut occupies about one half of the flat portion of the stick. The remaining half, or that more distant from the round section, is divided into two parts, which are separated by a rectangular space, in the centre of which there is a depression called the "*na-ta'l-tei*". On one side of the depression there are three notches, on the other, seven.

The stick found by Marshack matches this description quite well, and was probably collected by Fewkes at Hopi (see Zeilik, 1986a, for a photo).

Sticks or rocks (even rock art) with similar markings could be inferred to be Anasazi devices for tracking months, if found in the right archaeological context. More important, the mode of keeping such tallys would reveal the Anasazi solution to the problem of lunar calendar keeping and perhaps reveal whether or not the Anasazi paid attention to the lunar standstill cycle (Sofaer et al., 1982).

Summary

We have substantial evidence that the historic Pueblos used astronomical observations of the sun and moon to regulate the ritual and planting calendars. Special religious officials were empowered with the responsibility and authority to set the festival dates; they had to do so by forecasting the dates so that the proper preparations could be carried out. The anticipatory observations ensured that the ceremonies took place at the right time---on the days of the solstices, for example. The Pueblo people used two techniques to chart a seasonal calendar reliably and accurately: horizon profiles and interior projections of sunlight through special windows. Phases of the moon subdivided the seasonal calendar.

The Anasazi people confronted similar calendric problems for similar purposes and probably arrived at adaptive techniques similar to the historic Pueblos. Hence, the most appropriate strategy for exploring Anasazi astronomy in the proper cultural context involves hunting for material evidence of calendric skywatching in the prehistoric record.

Bibliography

Beaglehole, E. (1937). "Notes on Hopi Economic Life," Yale University Publications in Anthropology, 15.

Binford, L. R. (1972). "Methodological Considerations of the Archaeological Use of Ethnographic Data," in An Archaeological Perspective New York: Seminar Press, pp.59-67.

Bourke, J. G. (1881). "Diary," West Point, New York: United States Military Academy Library, Special Collections and Archives; photofacsimile in Special Collections, Coronado Room, Zimmerman Library, University of New Mexico. Entries for 1881.

Bunzel, R. L. (1932a). "Introduction to Zuni Ceremonialism," Bureau of American Ethnology 47th Annual Report, pp.467-544.

Bunzel, R. L. (1932b). "Zuni Ritual Poetry," Bureau of American Ethnology 47th Annual Report, pp.611-835.

Bunzel, R. L. (1932c). "Zuni Katcinas," Bureau of American Ethnology 47th Annual Report, pp.837-1086.

Carlson, J. B. (1987). "Romancing the Stone or Moonshine on the Sun Dagger," in Astronomy and Ceremony in the Prehistoric Southwest, eds. J. B. Carlson and W. J. Judge, Maxwell Museum Press: Albuquerque, Papers of the Maxwell Museum Press, No. 2, pp.71-88.

Cope, L. (1919). "Calendars of the Indians North of Mexico." University of California Publications in American Archaeology and Ethnology, vol. 16(4), pp.119-176.

Cordell, L. S. (1984). Prehistory of the Southwest, New York: Academic Press.

Cushing, F. H. (1920). Zuni Breadstuff. Museum of the American Indian, Heye Foundation, Indian Notes and Monographs 8; New York.

Cushing, F. H. (1979). "My Adventures in Zuni," reprinted in Zuni, ed. J. Green, Lincoln: University of Nebraska Press, pp.46-134.

Dutton, D. P. (1963). Sun Father's Way: The Kiva Murals of Kuaua, University of New Mexico Press: Albuquerque.

Ellis, F. H. (1975). "A Thousand Years of the Pueblo Sun-Moon-Star Calendar," in Archaeoastronomy in Precolumbian America, ed. A. F. Aveni, Austin: University of Texas Press, pp.59-87.

Fewkes, J. W. (1891). "A Few Summer Ceremonials at Zuni Pueblo," Journal of American Ethnology and Archaeology, i, pp.1-61.

Fewkes, J. W. (1892). "A Few Summer Ceremonials at the Tusayan Pueblos,"Journal of American Ethnology and Archaeology, ii, pp.1-161.

Fewkes, J. W. (1906). "Hopi Shrines Near East Mesa, Arizona," American Anthropologist (New Series), 8, pp.346-375.

Fewkes, J. W. (1919). "Prehistoric Villages, Castles, and Towers of Southwestern Colorado." Bureau of American Ethnology Bulletin No. 79, plates 14a, b, c; 19c and 32b.

Forde, C. D. (1931). "Hopi Agriculture and Land Ownership," Journal of the Royal Anthropological Institute of Great Britain and Ireland, 61, pp.357-405.

Frigout, E. (1979). "Hopi Ceremonialism," Handbook of North American Indians, vol. 9, ed. A. Ortiz, Washington: Smithsonian Institution, pp.564-576.

Harrington, J. P. (1916). "The Ethnography of the Tewa Indians," 29th Annual Report of the Bureau of American Ethnology, pp.29-36.

Hedges, K. (1985). "Methodology and Validity in California Archeoastronomy," in Earth and Sky, ed. A. Benson and M. Hoskinson, Slo'w Press: Thousand Oaks, pp.25-39.

Lange, C. H. (1959). Cochiti: A New Mexico Pueblo Past and Present, Austin: University of Texas Press.

Lange, C. H. and Riley, C. L. (1966). The Southwestern Journals of Adolph F. Bandelier: 1880-1882. Albuquerque: University of New Mexico Press.

Lyon, L. (1985). "Chronology of the Zuni Sha'lak'o Ceremony," in Southwestern Culture History: Collected Papers in Honor of Albert H. Schroeder, ed. C. H. Lange, Papers of the Archaeological Society of New Mexico, No. 10, pp.233-249.

McCluskey, S. C. (1977). "The Astronomy of the Hopi Indians," Journal for the History of Astronomy, viii, pp.174-195.

McCluskey, S. C. (1981). "Transformations of the Hopi Calendar," in Archaeoastronomy in the Americas, ed. R. A. Williamson, Ballena Press/Center for Archaeoastronomy, pp.173-182.

McCluskey, S. C. (1982). "Historical Astronomy: The Hopi Example," in Archaeoastronomy in the New World, ed. A. Aveni, Cambridge: Cambridge University Press, pp.31-57.

Newman, E. B., Mark, R. K., and Vivian, R. G. (1982). "Anasazi Solar Marker: The Use of a Natural Rockfall," Science, 217, pp.1036-1038.

Ortiz, A. (1969). The Tewa World, Chicago: University of Chicago Press.

Parsons, E. C. (1917). "Notes on Zuni," Memoirs of the American Anthropological Association, vol. 4(3-4), pp.151-327.

Parsons, E. C. (1925a). "A Pueblo Indian Journal 1920-1921," Memoirs of the American Anthropological Society, no. 32.

Parsons, E. C. (1939). Pueblo Indian Religion, Chicago: University of Chicago Press.

Preston, A. L. and Preston, R. A. (1985). "The Discovery of 19 Prehistoric Calendric Petroglyph Sites in Arizona," in Earth and Sky, ed. A. Benson and M. Hoskinson, Slo'w Press: Thousand Oaks, pp.123-133.

Preston, R. A. and Preston, A. L. (1987). "Evidence for the Calendric Function at 19 Prehistoric Petroglyph Sites in Arizona," in Astronomy and Ceremony in the Prehistoric Southwest, eds. J. B. Carlson and W. J. Judge, Maxwell Museum Press: Albuquerque, Papers of the Maxwell Museum of Anthropology, No. 2, pp.191-204.

Reyman, J. E. (1971). "Mexican Influence on Southwestern Ceremonialism,"Ph.D. dissertation, Southern Illinois University.

Reyman, J. E. (1975). "The Nature and Nurture of Archaeoastronomical Studies," in Archaeoastronomy in Precolumbian America, ed. A. F. Aveni, Austin: University of Texas Press, pp.205-215.

Reyman, J. E. (1976). "Astronomy, Architecture, and Adaptation at Pueblo Bonito," Science, 193, pp.957-962.

Reyman, J. E. (1980). "The Predictive Dimension of Priestly Power," in New Frontiers in Archaeology and Ethnohistory of the Greater Southwest, eds. C. L. Riley and B. C. Hendrick, Transactions of the Illinois Academy of Science, 72 (4), pp.40-59.

Schaafsma, P. (1985). "Form, Content, and Function: Theory and Method in North American Rock Art Studies," Advances in Archaeological Method and Theory, 8, pp.237-277.

Schroeder, A. H. (1979). "Pueblos Abandoned in Historic Times," in Southwest, ed. A. Ortiz, Handbook of the North American Indians, vol. 9, Washington: Smithsonian Institution, pp.236-254.

Sekaquaptewa, A. (1983). "Out of Phase with the Moon Phase," in Qua'toqi, Oraibi, Arizona

Simmons, L. W. (1942). Sun Chief: The Autobiography of a Hopi Indian, New Haven: Yale University Press.

Simmons, M. (1979). "History of Pueblo-Spanish Relations to 1821," in Southwest, ed. A. Ortiz, Handbook of the North American Indians, vol. 9, Washington: Smithsonian Institution, pp.178-193.

Sofaer, A. and Sinclair, R. (1987). "Astronomical Markings at Three Sites on Fajada Butte," in Astronomy and Ceremony in the Prehistoric Southwest, eds. J. B. Carlson and W. J. Judge, Maxwell Museum Press: Albuquerque, Papers of the Maxwell Museum of Anthropology, No. 2, pp.43-67.

Sofaer, A., Zinser, V., and Sinclair, R. M. (1979a). "A Unique Solar Marking Construct," in Science, 206, pp.283-291.

Sofaer, A. and Sinclair, R. M. (1986a), Letter, Science, 231, pp.1057-58.

Sofaer, A. and Sinclair, R. M. (1986b). Appraisal. Archaeoastronomy Supplement of the Journal for the History of Astronomy, no. 10, pp.559-566.

Sofaer, A., Sinclair, R. M., and Doggett, L. E. (1982). "Lunar Markings on Fajada Butte, Chaco Canyon, New Mexico," in Archaeoastronomy in the New World, ed. A. F. Aveni, Cambridge: Cambridge University Press, pp.169-181.

Spier, L. (1955). "Mohave Culture Items." Museum of Northern Arizona, Bulletin No. 28.

Stephen, A. M. (1936). Hopi Journal, ed. E. C. Parsons, New York: Columbia University Press.

Stevenson, M. C. (1904). "The Zuni Indians: Their Mythology, Esoteric Societies, and Ceremonies," Bureau of American Ethnology 23rd Annual Report.

Stevenson, M. C. (no date). material on the Tewa, Harrington Papers, Washington: Smithsonian Anthropological Archives.

Tedlock, B. (1983). "Zuni Sacred Theater," Native American Quarterly, 7, (4), p.93-110.

Titiev, M. (1938). "Dates of Planting at the Hopi Pueblo of Oraibi," Museum Notes of the Museum of Northern Arizona, xi:(5), pp.39-42.

Vierra, B. J. (1987). Test Excavations at Kuqua Pueblo (LA187). Laboratory of Anthropology Note No. 396.

White, L. A. (1942). "The Pueblo of Santa Ana, New Mexico," Memoirs of the American Anthropological Society, No. 60.

Williamson, R. A. (1979). "Field Report: Hovenweep National Monument." Archaeoastronomy, vol. 2 (3), pp.11-12.

Williamson, R. A. (1981). "North America: A Multiplicity of Astronomies," in Archaeology in the Americas, ed. R. A. Williamson, Ballena Press/Center for Archaeoastronomy, pp.61-80.

Williamson, R. A. (1982). "Casa Rinconada, Twelfth Century Anasazi Kiva," in Archaeoastronomy in the New World, ed. A. Aveni, Cambridge: Cambridge University Press, pp.205-218.

Williamson, R. A. (1983). "Sky Symbolism in a Navajo Rock Art Site, Chaco Canyon," Archaeoastronomy, vol. VI (1-4), pp.59-65.

Williamson, R. A. (1984). Living the Sky: The Cosmos of the American Indian, Boston: Houghton Mifflin.

Williamson, R. A. (1987). "Light and Shadow, Ritual, and Astronomy in Anasazi Structures," in Astronomy and Ceremony in the Prehistoric Southwest, eds. J. B. Carlson and W. J. Judge, Maxwell Museum Press: Albuquerque, Papers of the Maxwell Museum of Anthropology, No. 2, pp.99-119.

Williamson, R. A. and Young, M. J. (1979). "An Equinox Sun Petroglyph Panel at Hovenweep National Monument," American Indian Rock Art, vol. V, ed. F. G. Bock, K. Hedges, G. Lee, and H. Michaelis, El Toro, California, American Rock Art Research Association, pp.70-80.

Williamson, R. A., Fisher, H. J., Williamson, A. F., Cochran, C. (1975). "The Astronomical Record in Chaco Canyon, New Mexico," in Archaeoastronomy in Precolumbian America, ed. A. F. Aveni, Austin: University of Texas Press, pp.33-43.

Williamson, R. A., Fisher, H. J., and O'Flynn, D. (1977). "Anasazi Solar Observatories," in Native American Astronomy, ed. A. Aveni, Austin: University of Texas Press, pp.203-217.

Yava, A. (1978). Big Falling Snow, Albuquerque: University of New Mexico Press.

Young, M. J. (1986). "The Interrelationship of Rock Art and Astronomical Practice in the American Southwest," Archaeoastronomy Supplement to the Journal for the History of Astronomy, No. 10, pp.543-558.

Young, M. J. (1987a). "The Nature of the Evidence: Archaeoastronomy in the Prehistoric Southwest," in Astronomy and Ceremony in the Prehistoric Southwest, eds. J. B. Carlson and W. J. Judge, Maxwell Museum Press: Albuquerque, Papers of the Maxwell Museum of Anthropology, No. 2, pp.169-190.

Zeilik, M. (1980). "Pecked-Cross-Like Petroglyphs in New Mexico," Archaeoastronomy III (1), p.21.

Zeilik, M. (1984a). "A Possible Equinoctial Sun-Sighting Station at Tsiping, New Mexico," Archaeoastronomy, 7, (1-4), pp.70-75.

Zeilik, M. (1984b). "Summer Solstice at Casa Rinconada: Calendar, Hierophany, or Nothing?", Archaeoastronomy, 7, (1-4), pp.76-81.

Zeilik, M. (1985a). "The Ethnoastronomy of the Historic Pueblos. I. Calendrical Sun Watching," Archaeoastronomy Supplement to the Journal for the History of Astronomy, no. 8, pp.S1-S25.

Zeilik, M. (1985b). "The Fajada Butte Solar Marker: A Reevaluation," Science, 228, pp.1311-1313.

Zeilik, M. (1985c). "A Reassessment of the Fajada Butte Solar Marker," Archaeoastronomy Supplement to the Journal for the History of Astronomy, no. 9, pp.569-585.

Zeilik, M. (1985d). "Sun Shrines and Sun Symbols in the U. S. Southwest," Archaeoastronomy Supplement to the Journal for the History of Astronomy, no. 9, pp.586-596.

Zeilik, M. (1986a). "The Ethnoastronomy of the Historic Pueblos. II. Moon Watching", the Archaeoastronomy Supplement of the Journal for the History of Astronomy, No. 10, pp.S1-S22.

Zeilik, M. (1986b). "Keeping a Seasonal Calendar at Pueblo Bonito," Archaeoastronomy, vol. 8, pp.79-87.

Zeilik, M. (1986c). Reply. Science, 231, p.1058.

Zeilik, M. (1986d). Response. Archaeoastronomy Supplement of the Journal for the History of Astronomy, no. 10, pp.566-569.

Zeilik, M. (1987). "Anticipation in Ceremony: The Readiness is All," in Astronomy and Ceremony in the Prehistoric Southwest, eds. J. B. Carlson and W. J. Judge, Maxwell Museum Press: Albuquerque, Papers of the Maxwell Museum of Anthropology, No. 2, pp.25-41.

Zeilik, M. and Elston, R. (1983). "Wijiji at Chaco Canyon: A Winter Solstice Sunrise and Sunset Station," Archaeoastronomy, VI, (1-4), pp.66-73.

Zeilik, M. and Lowther,, M. A. (1988). "Calendric Sunwatching at Kuaua Pueblo," paper presented at the 171st meeting of the American Astronomical Society, Austin, Jan. 1988.

www.ingramcontent.com/pod-product-compliance
Lightning Source LLC
LaVergne TN
LVHW071208210926
848482LV00015B/756